Eugene E. Tyrtyshnikov

A Brief Introduction
to Numerical Analysis

Birkhäuser
Boston • Basel • Berlin

Eugene E. Tyrtyshnikov
Institute of Numerical Mathematics
Russian Academy of Sciences
Leninski Prospekt 32A
117 334 Moscow
Russia

Library of Congress Cataloging-in-Publication Data

Tyrtyshnikov, E. E. (Evgeniĭ Evgen'evich)
    A brief introduction to numerical analysis / Eugene Tyrtyshnikov.
        p.   cm.
    Includes bibliographical references and index.
    ISBN 0-8176-3916-0 (hc : alk. paper). -- ISBN 3-7643-3916-0 (alk.
paper)
    1. Numerical analysis.   I. Title.
QA297.T97   1997
519.4--dc21   97-301
                                                         CIP

Printed on acid-free paper
© 1997 Birkhäuser Boston   *Birkhäuser*

ISBN 0-8176-3916-0
ISBN 3-7643-3916-0
Typeset by the Author in LATEX.
Printed and bound by Maple Vail, York, PA.
Printed in the U.S.A.

9 8 7 6 5 4 3 2 1

Preface

Probably I ought to explain why one more book on numerical methods can be useful. Without any doubt, there are many quite good and excellent books on the subject. But I know definitely that I did not realize this when I was a student. In this book, my first desire was to present those lectures that I wished I would have heard when I was a student.

Besides, in spite of the profusion of textbooks, introductory courses, and monographs on numerical methods, some of them are too elementary, some are too difficult, some are far too overwhelmed with applications, and most of them are too lengthy for those who want to see the whole picture in a short time.

I hope that the brevity of the course left me no chance to obscure the beauty and depth of mathematical ideas behind the theory and methods of numerical analysis.

I am convinced that such a book should be very concise indeed. It should be thoroughly structured, giving information in short sections which, ideally, are a half-page in length. Equally important, the book should not give an impression that nothing is left to work on in this field. Any time it becomes possible to say something about modern development and recent results, I do try to find time and place for this.

Still, I do not promise *easy* reading. This book is addressed, first, to those who study mathematics. Despite this, it is written so that it can be read by students majoring in physics and mathematics, and I believe it can be useful for advanced readers and researchers providing them with new findings and a new vision of the basic framework.

Somebody might remark that there is no excuse for brevity in the list of references at the end of this book. I could only agree and beg not to be blamed for this. I included in the list only the books that I felt influenced me most directly. Several imposing papers are also mentioned in the footnotes.

The book contains, in fact, a concise and closed exposition of the lectures given by the author to the 2–3 year students of the Chair of Mathematical Modelling of Physical Processes of the Faculty of Problems of Physics and Energetics of the Moscow Institute of Physics and Technology.

To conclude the preface, I get to its main purpose, to express my thanks. Above all, I am grateful to V. V. Voevodin, my first teacher who had inspired me by his way of doing science. His advice and encouragement were always of great importance to me.

Special thanks go to S. A. Goreinov and N. L. Zamarashkin. They were the first readers and found many opportunities to share with me their remarks and impressions.

It is my pleasure also to express my gratitude to G. I. Marchuk for suggesting these lectures.

December 1996   Eugene Tyrtyshnikov

Contents

Lecture 1

1.1 Metric space

For various mathematical objects, we want to become aware how to compute them. However, any algorithm produces no more than some other objects which, as we hope, are "close" to the ones in question. Thus, we need to have a rigorous definition of "closeness" for different objects.

Generally, this can be done with the help of a "metric" (or "distance"). Let M be a nonempty set and let $\rho(x, y)$ be a nonnegative function defined for all $x, y \in M$ and enjoying the following properties:

(1) $\rho(x, y) \geq 0 \quad \forall x, y \in M$;
$\rho(x, y) = 0 \quad \Leftrightarrow \quad x = y$;

(2) $\rho(x, y) = \rho(y, x) \quad \forall x, y \in M \quad$ (symmetry);

(3) $\rho(x, y) \leq \rho(x, y) + \rho(y, z) \quad \forall x, y, z \in M \quad$ (triangle inequality).

Such a function $\rho(x, y)$ is called the metric, or distance (between x and y), and M is termed the metric space in this case.

A very familiar example of the metric space: M is the set of all real numbers and $\rho(x, y) \equiv |x - y|$.

Another instructive example: M is an arbitrary nonempty set; $\rho(x, y) = 0$ for $x = y$ and 1 for $x \neq y$.

1.2 Some useful definitions

A sequence $x_n \in M$ is called *convergent* if $\exists x \in M : \lim_{n \to \infty} \rho(x_n, x) = 0$. It is easy to prove that such a point x is always unique; x is called the limit for x_n. The notation: $x = \lim_{n \to \infty} x_n$.

A sequence $x_n \in M$ is said to be *Cauchy sequence* if

$$\forall \varepsilon > 0 \quad \exists N : \quad n, m \geq N \Rightarrow \rho(x_n, x_m) \leq \varepsilon.$$

A metric space M is termed *complete* if any Cauchy sequence in it is convergent.

A set $C \subset M$ is referred to as *closed* if for any convergent sequence $x_n \in C$ its limit belongs to C.

A closed set $C \subset M$ is called *compact* if for any sequence in it there is a subsequence that is convergent.

A set $B(a;r) \equiv \{x \in M : \rho(x,a) < r\}$ is referred to as an *open ball* with the center at point a and radius r. A set $\overline{B}(a;r) \equiv \{x \in M : \rho(x,a) \le r\}$ is called a *closed ball*.

A set $O \subset M$ is called *open* if any point x is included therein together with some open ball $B(x,r)$.

A set $S \subset M$ is said to be *bounded* if the whole of it belongs to some ball.

A set S_1 is *dense* in a set S_2 if for every point $x \in S_2$ there is a sequence of points from S_1 convergent to x.

A numeric function $f(x), x \in M$, is termed *continuous* at a point x_0 if for any sequence $x_n \ne x_0$ such that $x_0 = \lim_{n \to 0} x_n$, $f(x_0) = \lim_{n \to \infty} f(x_n)$.

1.3 Nested balls

Theorem 1.3.1 *Given closed balls $\overline{B}(a_n, r_n)$ in a complete metric space M, assume that*

(1) $\overline{B}(a_1;r_1) \supset \overline{B}(a_2;r_2) \supset \dots$;

(2) $\lim_{n \to \infty} r_n = 0$.

Then the intersection of all these balls $P = \bigcap\limits_{n=1}^{\infty} \overline{B}(a_n, r_n)$ is not empty and contains exactly one point.

The requirement (2) is important. To show this, consider an "exotic" metric space as follows: $M = \{1, 2, \dots\}$, and

$$\rho(m,n) = \begin{cases} 0, & m = n; \\ 1 + \max\left(\frac{1}{2^m}, \frac{1}{2^n}\right), & m \ne n. \end{cases}$$

It is not difficult to verify that ρ is a metric indeed (the properties (1)–(3) in Section 1.1 are fulfilled). Moreover, we have a complete metric space. At the same time, the balls

$$\overline{B}\left(1, 1 + \tfrac{1}{2}\right) \supset \overline{B}\left(2, 1 + \tfrac{1}{2^2}\right) \supset \overline{B}\left(3, 1 + \tfrac{1}{2^3}\right) \supset \dots$$

have no point in common.

1.4 Normed space

Assume that V is a real or complex vector space on which a nonnegative function $f(x)$ is defined so that

(1) $f(x) \geq 0 \quad \forall x \in V; \quad f(x) = 0 \Leftrightarrow x = 0;$

(2) $f(\alpha x) = |\alpha| f(x), \quad \alpha \in \mathbb{R} \quad (\text{or} \quad \mathbb{C}), \quad x \in V;$

(2) $f(x + y) \leq f(x) + f(y) \quad$ (the triangle inequality).

Such a function $f(x)$ is called the norm of the vector x while V is called the normed space. The notation: $\|x\| \equiv f(x)$.

For any normed space, the metric is introduced as follows: $\rho(x, y) \equiv \|x - y\|$. The convergence and other notions considered in Section 1.2 are understood in the sense of this metric.

A complete normed space is called the *Banach space.*

1.5 Popular vector norms

Let $V = \mathbb{C}^n$ (or $\mathbb{R}^n$). If $p \geq 1$ and $x = [x_1, \ldots, x_n]^T$, then assume that

$$\|x\|_p \equiv \left(\sum_{i=1}^{n} |x_i|^p \right)^{1/p} \quad \text{(the p-norm of x).}$$

Theorem 1.5.1 $\|x\|_p$ *is a norm.*

The properties (1) and (2) of the norm are evident. The property (3) is given by the Minkowski inequality, which will be proven below.

Lemma 1.5.1 *Assume that numbers p, q make up the Hölder pair, i.e.,*

$$p, q \geq 1; \quad \frac{1}{p} + \frac{1}{q} = 1.$$

Then for all $a, b \geq 0$,

$$ab \leq \frac{a^p}{p} + \frac{b^q}{q}.$$

We can prove this easily using the concavity of the logarithmic function. The concavity means that for all $u, v > 0$,

$$\alpha \log u + \beta \log v \leq \log (\alpha u + \beta v),$$

$$\forall \alpha, \beta \geq 0, \quad \alpha + \beta = 1.$$

Theorem 1.5.2 (The Hölder inequality). *Assume that p, q are an arbitrary Hölder pair. Then for any vectors $x = [x_1, \ldots, x_n]^T$, $y = [y_1, \ldots, y_n]^T$,*

$$\left| \sum_{i=1}^{n} x_i y_i \right| \leq \|x\|_p \|y\|_q.$$

Proof. If $x = 0$ and $y = 0$, then the inequality is trivial. Therefore, consider nonzero vectors x, y and set

$$\tilde{x} = x/\|x\|_p, \qquad \tilde{y} = y/\|y\|_q.$$

Then $\|\tilde{x}\|_p = \|\tilde{y}\|_q = 1$. According to Lemma 1.5.1,

$$|\tilde{x}_i| \, |\tilde{y}_i| \le \frac{|\tilde{x}_i|^p}{p} + \frac{|\tilde{y}_i|^q}{q}, \quad i = 1, \dots, n.$$

Adding up these inequalities we obtain

$$\sum_{i=1}^{n} |\tilde{x}_i| \, |\tilde{y}_i| \le \frac{\|\tilde{x}\|_p^p}{p} + \frac{\|\tilde{y}\|_q^q}{q} = 1. \ \square$$

Theorem 1.5.3 (The Minkowski inequality)

$$\|x + y\|_p \le \|x\|_p + \|y\|_p.$$

Proof.

$$\|x + y\|_p^p = \sum_{i=1}^{n} |x_i + y_i|^p \le \sum_{i=1}^{n} |x_i + y_i|^{p-1} (|x_i| + |y_i|)$$

(next we use the Hölder inequality)

$$\le \left(\sum_{i=1}^{n} \left(|x_i + y_i|^{p-1} \right)^q \right)^{\frac{1}{q}} (\|x\|_p + \|y\|_p).$$

It remains to recall that $(p - 1)q = p$. $\square$

The following norm is also regarded as a p-norm:

$$\|x\|_\infty \equiv \max_i |x_i|.$$

It is easy to prove that this is a norm, indeed, and that

$$\|x\|_\infty = \lim_{p \to \infty} \|x\|_p.$$

The p-norms with $p = 1, 2$ and ∞ are most widespread. Here is what the unit spheres for these norms look like $(n = 2)$:

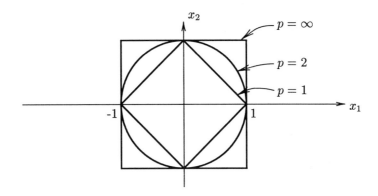

Some people call the 1-norm octahedral and the ∞-norm cubic. Can you explain why?

1.6 Matrix norms

If we take up all matrices of the same size, we can treat them as a finite-dimensional vector space. Consequently, a norm for matrices can be brought in through any vector norm. By a matrix norm, however, we mean something more than this.

Assume that any matrix A is endowed with a number $\|A\|$. Then $\|A\|$ is called the matrix norm if

(1) $\|A\|$ is a vector norm on any space of matrices of the same size;

(2) for any matrices A and B that can be multiplied,

$$\|AB\| \leq \|A\|\|B\| \quad \text{(the submultiplicative property)}.$$

One of the most important examples of the matrix norm is the *Frobenius norm*:

$$\|A\|_F \equiv \left(\sum_{i=1}^{m} \sum_{j=1}^{n} |a_{ij}|^2 \right)^{1/2}, \qquad A \in \mathbb{C}^{m \times n}.$$

Proposition. *The Frobenius norm possesses the submultiplicative property.*

Proof. Let

$$A = [a_1, \ldots, a_n] \in \mathbb{C}^{m \times n}, \qquad B = \begin{bmatrix} b_1^T \\ \ldots \\ b_n^T \end{bmatrix} \in \mathbb{C}^{n \times k}.$$

Then
$$AB = a_1 b_1^T + \cdots + a_n b_n^T.$$

From the triangle inequality,

$$
\begin{aligned}
\|AB\|_F &\leq \|a_1 b_1^T\|_F + \cdots + \|a_n b_n^T\|_F \\
&= \|a_1\|_2 \|b_1\|_2 + \cdots + \|a_n\|_2 \|b_n\|_2 \\
&\leq \left(\sum_{i=1}^n \|a_i\|_2^2 \right)^{\frac{1}{2}} \left(\sum_{i=1}^n \|b_i\|_2^2 \right)^{\frac{1}{2}} = \|A\|_F \|B\|_F. \quad \square
\end{aligned}
$$

1.7   Equivalent norms

Norms $\|\cdot\|_*$ and $\|\cdot\|_{**}$ on the same vector space V are called *equivalent* if there exist $c_1, c_2 > 0$ such that

$$c_1 \|x\|_* \leq \|x\|_{**} \leq c_2 \|x\|_* \qquad \forall\, x \in V.$$

Clearly, the equivalent norms are of equal worth from the standpoint of convergence. The next theorem is the fundamental fact valid for finite-dimensional spaces.

Theorem 1.7.1 *Any two norms on a finite-dimensional space are equivalent.*

Proof. First of all, we need the following basic facts:

1) The compactness of the unit sphere $S_n = \{x \in \mathbb{R}^n : \|x\|_2 = 1\}$ with respect to the 2-norm.

2) The continuity of any norm $\|\cdot\|_*$ with respect to the 2-norm.

3) The Weierstrass theorem stating that any function continuous on a compact set is bounded.

The compactness of the unit sphere. Consider a sequence

$$x^{(k)} = [x_1^{(k)}, \ldots, x_n^{(k)}]^T \in S_n.$$

The sequence of the first coordinates $x_1^{(k)}$ belongs to the interval $[-1, 1]$, and hence it possesses a convergent subsequence: $x_1^{(k_1)} \to x_1$. Consider the subsequence $x^{(k_1)}$ and the second coordinates $x_2^{(k_1)}$. Let $x_2^{(k_2)} \to x_2$. Further, consider the subsequence $x^{(k_2)}$, the third coordinates $x_3^{(k_2)}$, and so on. In the end we shall have a sequence of the vectors $x^{(k_n)}$ such that all its coordinate sequences are convergent: $x_i^{(k_n)} \to x_i$.

Thus, if $x \equiv [x_1, \ldots, x_n]^T$, then

$$\left\| x^{(k_n)} - x \right\|_2 = \left(\sum_{i=1}^n \left| x_i^{(k_n)} - x_i \right|^2 \right)^{\frac{1}{2}} \to 0.$$

The continuity of a norm. Assume that $x^{(k)} \to x$, i.e., $\left\| x^{(k)} - x \right\|_2 \to 0$. We want to prove that $\left\| x^{(k)} \right\|_* \to \|x\|_*$. Denote by $e_1, \ldots, e_n$ the columns of the unity matrix. Then

$$
\begin{aligned}
\left| \left\| x^{(k)} \right\|_* - \|x\|_* \right| &\leq \left\| x^{(k)} - x \right\|_* \\
&= \left\| \sum_{i=1}^n \left(x_i^{(k)} - x_i \right) e_i \right\|_* \leq \sum_{i=1}^n \left| x_i^{(k)} - x_i \right| \|e_i\|_* \\
&\leq \left\| x^{(k)} - x \right\|_2 \left(\sum_{i=1}^n \|e_i\|_*^2 \right)^{\frac{1}{2}} \to 0.
\end{aligned}
$$

The Weierstrass theorem. Assume that M is a compact set and a numeric function $f(x)$ is continuous at any of its points. Assume that $f(x)$ is not bounded. Then there exists a subsequence $x^{(k)}$ such that $\left| f(x^{(k)}) \right| \geq k$. Due to the compactness property there is a subsequence that converges: $x^{k'} \to x$. From the continuity, $f\left(x^{(k')} \right) \to f(x)$. However, this cannot be true because

$$
k \leq \left| f\left(x^{(k')} \right) \right| \leq |f(x)| + \left| f\left(x^{(k')} \right) - f(x) \right|
$$

for all k. Hence, the function $f(x)$ is bounded.

Now, the function $\|x\|_*$ is continuous on the compact set S_n with respect to the 2-norm and thence is bounded, i.e., for some $c_2 > 0$ $\|x\|_* \leq c_2$. The function $1 / \|x\|_*$ is continuous on S_n as well; hence, for some $c_1 > 0$ $1 / \|x\|_* \leq c_1^{-1}$. Therefore, for all $x \in S_n$,

$$
c_1 \leq \|x\|_* \leq c_2.
$$

If $x \notin S_n$, $x \neq 0$, then $x / \|x\|_2 \in S_n$. Thus,

$$
c_1 \|x\|_2 \leq \|x\|_* \leq c_2 \|x\|_2.
$$

We have proven that the norm $\| \cdot \|_*$ is equivalent to the norm $\| \cdot \|_2$. It follows immediately that $\| \cdot \|_*$ is equivalent to any other norm. □

1.8   Operator norms

Let the norm $\| \cdot \|_*$ be defined on $\mathbb{C}^m$ while $\| \cdot \|_{**}$ on $\mathbb{C}^n$. Then for $A \in \mathbb{C}^{m \times n}$ we set

$$
\|A\|_{***} = \max_{x \neq 0} \frac{\|Ax\|_*}{\|x\|_{**}}.
$$

Prove that the maximum exists![1]

---

[1] Remember the results from the previous section.

It is not difficult to check that $\|A\|_{***}$ is an operator norm on $\mathbb{C}^{m \times n}$. It is called the operator norm induced by the vector norms $\| \cdot \|_*$ and $\| \cdot \|_{**}$.

For an operator norm, the following *compatibility property* holds:

$$\|Ax\|_* \le \|A\|_{***}\|x\|_{**},$$

which obviously follows from the definition of $\|A\|_{***}$.

By definition,

$$\|A\|_p \equiv \max_{x \ne 0} \frac{\|Ax\|_p}{\|x\|_p}.$$

This is an operator norm. Actually, if $AB \ne 0$ then

$$
\begin{aligned}
\|AB\|_p &= \max_{x \ne 0, Bx \ne 0} \frac{\|ABx\|_p}{\|Bx\|_p} \frac{\|Bx\|_p}{\|x\|_p} \\
&\le \max_{x \ne 0, Bx \ne 0} \frac{\|ABx\|_p}{\|Bx\|_p} \cdot \max_{x \ne 0} \frac{\|Bx\|_p}{\|x\|_p} \le \|A\|_p\|B\|_p. \quad \square
\end{aligned}
$$

Pay attention to some useful formulas (prove them!):

$$
\|A\|_1 = \max_{1 \le j \le n} \sum_{i=1}^{m} |a_{ij}|,
$$

$$
\|A\|_\infty = \max_{1 \le i \le n} \sum_{j=1}^{n} |a_{ij}|, \qquad A \in \mathbb{C}^{m \times n}.
$$

We shall see shortly that $\|A\|_2^2$ is the maximal eigenvalue of the matrix A^*A. That is why the operator norm $\|A\|_2$ is often referred to as the *spectral norm*.

Exercises

1. Devise a metric for which the set of all real numbers is not a Banach space.

2. Consider a sequence of nested closed balls in a Banach space. Prove that if their radii tend to zero, then all the balls have exactly one point in common.

3. Show that a sequence of open nested balls can have an empty intersection, even if their radii tend to zero.

4. A norm is called absolute if $\|x\| = \| \, |x| \, \|$, where $|x|$ is the vector made up of the absolute values of the components of the vector x. Produce a norm that is not absolute.

5. Prove that the operator norm is a norm.

6. Prove the formulas for $||A||_1$ and $||A||_\infty$ from Section 1.6.

7. Let the norm $|| \cdot ||$ be given on $\mathbb{C}^n$. The operator norm

$$||x||_* = \max_{y \neq 0} \frac{|x^T y|}{||y||}, x \in \mathbb{C}^n,$$

is called a *dual norm* for $|| \cdot ||$. Prove that for the p-norm, the dual norm is the q-norm, where p and q are the Hölder pair.

8. Suppose that a matrix $A \in \mathbb{C}^{n \times n}$ preserves the p-norm:

$$||Ax||_p = ||x||_p \quad \forall x \in \mathbb{C}^{n \times n}.$$

Prove that this holds if and only if A^T preserves the q-norm:

$$||A^T x||_q = ||x||_q \quad \forall x \in \mathbb{C}^{n \times n}$$

(p and q are the Hölder pair).

9. Prove that the Frobenius norm cannot be an operator norm.

10. P. Groen built up an example of a submultiplicative matrix norm that takes the value 1 at the unity matrix but is not an operator norm:

$$||A||_{G(c)} \equiv \max_{1 \leq i \leq n} \left(|a_{ii}| + c \sum_{j \neq i} |a_{ij}| \right), \quad c > 1,$$

$$A = [a_{ij}] \in \mathbb{C}^{n \times n}.$$

Prove it!

11. Produce an example of norms that are not equivalent.

12. Prove that the ball $B = \overline{B}(0; 1)$ for a norm on $\mathbb{R}^n$ possesses the following properties:

(1) B is a compact set with respect th the 2-norm;
(2) if $x, y \in B$ and $0 \leq \alpha \leq 1$, then $\alpha x + (1 - \alpha)y \in B$ (concavity);
(3) if $x \in B$ and $|\alpha| \leq 1$, then $\alpha x \in B$;
(4) $\exists r > 0 : \{y : ||y||_2 < r\} \subset B$.

Prove that for an arbitrary set $B \subset \mathbb{R}^n$ enjoying the properties (1)–(4), there exists a norm for which

$$B = \overline{B}(0, 1).$$

13. Prove that if A is a submatrix in B, then $||A||_p \leq ||B||_p$.

14. Can the norm of a submatrix be greater than the norm of the whole matrix?

15. The elements of A and B are nonnegative, and $a_{ij} \leq b_{ij}$ for all i, j. Is it true that $\|A\|_p \leq \|B\|_p$?

16. Suppose $f(A)$ is a norm on $\mathbb{C}^{n \times n}$ not necessary enjoying the submultiplicative property. Prove that a number c exists such that $\|A\| \equiv c\,f(A)$ acquires the latter property.

Lecture 2

2.1 Scalar product

Assume that V is a real or complex vector space on which, for any pair of vectors x and y, a number (x, y) is defined so that

(1) $(x, x) \geq 0 \quad \forall x$; $(x, x) = 0 \Leftrightarrow x = 0$;

(2) $(x, y) = \overline{(y, x)}$;

(3) $(\alpha x, y) = \alpha(x, y)$, α is a number;

(4) $(x + y, z) = (x, z) + (y, z)$.

Then (x, y) is called the *scalar product* of the vectors x and y.

A real space with a scalar product is called *Euclidean*. A complex space with a scalar product is called *unitary*.

If $(x, y) = 0$, then the vectors are called *orthogonal*. Assume that V is of dimension n, and let x_i and y_i be the coordinates of the vectors x and y in their expansions in some basis of V. If $(x, y) = x_1 \bar{y}_1 + \ldots + x_n \bar{y}_n$ for all x and y, then the basis is called *orthonormal*.

2.2 Length of a vector

Given a scalar product, we can naturally define the *length* of a vector x as $(x, x)^{1/2}$. Prove that the length of a vector is a vector norm. It is talked about as the norm induced by the scalar product.

If $\|x\| \equiv (x, x)^{1/2}$, then the *parallelogram identity* is true:

$$\|x + y\|^2 + \|x - y\|^2 = 2\|x\|^2 + 2\|y\|^2.$$

It follows immediately that there are some norms not induced by any scalar product (for instance, $\|x\|_1$).

Theorem 2.2.1 *A norm on a vector space is induced by a scalar product if and only if it is subject to the parallelogram identity.*

Proof. For simplicity, consider a real space. Set

$$(x, y) \equiv \tfrac{1}{2} \left(\|x + y\|^2 - \|x\|^2 - \|y\|^2 \right),$$

and try to prove that this is a scalar product. Properties (1) and (2) in Section 2.1 are evident. Property (4) is equivalent to the following identity:

$$\|x + y + z\|^2 - \|x + y\|^2 - \|z\|^2$$
$$= \left(\|x + z\|^2 - \|x\|^2 - \|z\|^2 \right) + \left(\|y + z\|^2 - \|y\|^2 - \|z\|^2 \right). \tag{$*$}$$

To derive it, we apply the parallelogram identity twice:

$$\|x + y + 2z\|^2 = \|(x + y + z) + z\|^2 = 2\|x + y + z\|^2 + 2\|z\|^2 - \|x + y\|^2;$$
$$\|x + y + 2z\|^2 = \|(x + z) + (y + z)\|^2 = 2\|x + z\|^2 + 2\|y + z\|^2 - \|x - y\|^2.$$

From these two equations we find that

$$\|x + y + z\|^2 = \tfrac{1}{2}\|x + y + 2z\|^2 - \|z\|^2 + \tfrac{1}{2}\|x + y\|^2;$$
$$\|x + z\|^2 + \|y + z\|^2 = \tfrac{1}{2}\|x + y + 2z\|^2 + \tfrac{1}{2}\|x - y\|^2.$$

We substitute the first equation in the left-hand side and the second in the right-hand side of $(*)$. Once more, remembering the parallelogram identity, we see that the both sides coincide.

If α is rational, then the property (3) follows from (4). Thanks to the continuity with respect to α, it holds for all real α. $\qquad\square$

2.3   Isometric matrices

A matrix $Q \in \mathbb{C}^{n \times n}$ is called *norm-preserving* $\| \cdot \|$ on $\mathbb{C}^n$, or *isometric* in the norm $\| \cdot \|$ on $\mathbb{C}^n$, if

$$\|Qx\| = \|x\| \quad \forall\, x \in \mathbb{C}^n.$$

What can be said about the matrices that preserve the p-norms? As a simple example, consider matrices that can be obtained from the unity matrix by swapping rows (or columns). Such matrices are known as *permutational matrices*. Apart from $p = 2$, isometric matrices are not too different from permutational matrices.

Consider as x the columns of the unity matrix $\Rightarrow$ the p-norm of each column of Q is equal to 1. Moreover, if p and q are the Hölder pair, then

$$
\begin{aligned}
\|Q^T y\|_q &= \max_{x \neq 0} \frac{|y^T Q x|}{\|x\|_p} = \max_{x \neq 0} \frac{|y^T Q x|}{\|Qx\|_p} \\
&= \max_{z \neq 0} \frac{|y^T z|}{\|z\|_p} = \|y\|_q.
\end{aligned}
$$

It follows that the q-norm of each row of Q is equal to 1.

Let $p < 2$. Then the 2-norm of each column is at least 1, while the 2-norm of each row is at most $1 \Rightarrow$ the 2-norm of each column and each row is equal to 1. If a column has more than one nonzero components than they are less than 1 in modulus $\Rightarrow$ the p-norm of such a column must be less than 1. To dismiss the contradiction, we conclude that in each column of Q there is exactly one element equal to 1 in modulus. In the corresponding row, all other elements are equal to 0. (The case $p > 2$ is treated similarly.)

Thus, if $p \neq 2$, then the matrix Q preserving the p-norm is of the form

$$Q = P \, \mathrm{diag}(d_1, \ldots, d_n),$$

where P is the permutational matrix and $|d_i| = 1, i = 1, \ldots, n$.

2.4 Preservation of length and unitary matrices

For $p = 2$, the set of isometric matrices is much richer. In this case, the preservation of the 2-norm implies the preservation of the scalar product (prove this!) $\Rightarrow$ the columns $q_1, \ldots, q_n$ of Q comprise an orthonormal system:

$$q_i^* q_j = \delta_{ij} \Leftrightarrow Q^* Q = I$$

(δ is the Kronecker symbol: $\delta_{ij} = 1$ if $i = j$ and 0 if $i \neq j$).

A matrix $Q \in \mathbb{C}^{n \times n}$ such that $Q^* = Q^{-1}$ is called *unitary*.

The unitary matrices excel in being the only ones that preserve the length (2-norm) and the scalar product.

The important property of unitary matrices: their products and inverses remain unitary (prove this).

2.5 Schur theorem

Theorem 2.5.1 (Schur) *For any matrix $A \in \mathbb{C}^{n \times n}$ with the eigenvalues $\lambda_1, \ldots, \lambda_n$, there exists a unitary matrix Q such that*

(1) *the matrix $Q^* A Q$ is upper triangular,*

(2) $\mathrm{diag}(Q^* A Q) = \mathrm{diag}(\lambda_1, \ldots, \lambda_n)$.

Proof. Assume that $Av_1 = \lambda_1 v_1, \|v_1\|_2 = 1$ and choose $v_2, \ldots, v_n$ so that the matrix $V_1 = [v_1, v_2, \ldots, v_n]$ is unitary. Then

$$V_1^* A V_1 = \begin{bmatrix} \lambda_1 & * & \cdots & * \\ 0 & & & \\ \cdots & & A_1 & \\ 0 & & & \end{bmatrix}.$$

Proceed by induction. $\square$

2.6   Normal matrices

A matrix A is called *normal* if $A^*A = AA^*$.

The most important classes of normal matrices:

(1) *Hermitian* matrices: $H^* = H$;

(2) *unitary* matrices: $U^* = U^{-1}$.

In the case of real elements, a Hermitian matrix is called symmetric and a unitary matrix is called orthogonal.

Theorem 2.6.1 *A matrix $A \in \mathbb{C}^{n \times n}$ is normal if and only if $\mathbb{C}^n$ has an orthonormal basis of its eigenvectors.*

Proof. For any $A \in \mathbb{C}^{n \times n}$, a unitary matrix U exists such that $T = U^*AU$ is an upper triangular matrix (the Schur theorem). Further, $A^*A = AA^*$ is equivalent to $T^*T = TT^*$, and, as easily seen, any upper triangular matrix with such a property is bound to be diagonal. Thus, the columns of U yield a basis of the eigenvectors of A. $\square$

Theorem 2.6.2 *A normal matrix A is Hermitian if and only if its eigenvalues are all real.*

Theorem 2.6.3 *A normal matrix is unitary if and only if its eigenvalues are all equal to 1 in modulus.*

Prove these theorems.

It might be useful to keep in mind that an arbitrary matrix $A \in \mathbb{C}^{n \times n}$ can be split (uniquely) as

$$A = H + \mathrm{i}\, K, \quad H^* = H, \quad K^* = K, \quad \mathrm{i}^2 = 1.$$

This is the so-called *Hermitian decomposition* of A.

It is trivial to prove that for A to be normal, it is necessary and sufficient that H and K commute.

2.7   Positive definite matrices

Among Hermitian matrices, we distinguish those for which the scalar product $(Ax, x) = x^*Ax$ keeps the same sign for all x.

In the case $(Ax, x) \geq 0 \quad \forall x \in \mathbb{C}^{n \times n}$, the matrix A is called *positive semidefinite*, or *nonnegative definite*. The notation: $A \geq 0$.

In the case $(Ax, x) > 0 \quad \forall x \in \mathbb{C}^{n \times n}$, the matrix A is called *positive definite*. The notation: $A > 0$.

Definiteness implies Hermitianness. For $A \geq 0$, consider its Hermitian decomposition $A = H + \mathrm{i}\, K$. Since

$$(Ax, x) = (Hx, x) + \mathrm{i}\,(Kx, x) \in \mathbb{R} \quad \forall x \in \mathbb{C}^n,$$

$(Kx, x) = 0 \quad \forall x \in \mathbb{C}^n \Rightarrow$ all the eigenvalues of the Hermitian matrix K are equal to 0 $\Rightarrow$ $K = 0$. $\square$

Note that if $A \in \mathbb{R}^{n \times n}$ and $(Ax, x) \geq 0 \quad \forall x \in \mathbb{R}^n$, then the matrix A is not bound to be symmetric.

For $A \in \mathbb{C}^{n \times n}$ to be nonnegative (positive) definite, it is necessary and sufficient that all its eigenvalues be nonnegative (positive). Prove this.

The important property: a matrix is nonnegative (positive) definite if and only if all its leading submatrices are the same. (A submatrix is called *leading* if it occupies the upper-left corner of the matrix.) To prove this, it is sufficient to observe that

$$[y^* \quad 0] \begin{bmatrix} B & * \\ * & * \end{bmatrix} \begin{bmatrix} y \\ 0 \end{bmatrix} = y^* B y \quad \forall y.$$

Prove that $A^* A \geq 0$ for an arbitrary A (it is simple). We are going to use this no later than in the next section.

2.8 The singular value decomposition

Theorem 2.8.1 *Suppose $A \in \mathbb{C}^{m \times n}$, $r = \operatorname{rank} A$. Then there exist positive numbers $\sigma_1 \geq \ldots \geq \sigma_r > 0$ and unitary matrices $U \in \mathbb{C}^{n \times n}$, $V \in \mathbb{C}^{m \times m}$ for which there holds*

$$A = V \Sigma U^*, \tag{2.8.1}$$

where

$$\Sigma = \begin{bmatrix} \sigma_1 & & \\ & \ddots & 0 \\ 0 & & \sigma_r \\ & & \end{bmatrix} \in \mathbb{C}^{m \times n}. \tag{2.8.2}$$

Proof. $A^* A \geq 0 \Rightarrow$ there exists a unitary matrix $U = [u_1, \ldots, u_n] \in \mathbb{C}^{n \times n}$ such that

$$U^* A^* A U = \operatorname{diag}(\sigma_1^2, \ldots, \sigma_n^2).$$

Suppose $\sigma_i > 0$ for $1 \geq i \geq r$ and $\sigma_i = 0$ for $i > r$. Let $U_r = [u_1, \ldots, u_r]$ and $\Sigma_r = \operatorname{diag}(\sigma_1, \ldots, \sigma_r)$. Then

$$U_r^* A^* A\, U_r = \Sigma_r^2 \quad \Rightarrow \quad (\Sigma_r^{-1} U_r^* A^*)(A U_r \Sigma_r^{-1}) = I.$$

Hence, the matrix $V_r = A U_r \Sigma_r^{-1}$ is such that $V_r^* V_r = I$ (V_r has orthonormal columns) $\Rightarrow$

$$V_r^* A U_r = \Sigma_r.$$

Embedding V_r, in any way, into a unitary matrix $V \in \mathbb{C}^{m \times m}$ and taking into account the formula for V_r, we obtain

$$V^* A U = \begin{bmatrix} \Sigma_r & 0 \\ 0 & 0 \end{bmatrix}. \quad \square$$

Decomposition (2.8.1) is called the *singular value decomposition* (SVD) of the matrix A. The numbers $\sigma_1 \geq \ldots \geq \sigma_r > 0$ are called the singular values while the vectors u_i are the right and v_i the left singular vectors of A. It is customary to say that, apart from having r nonzero singular values, the matrix A has yet $\min(m, n) - r$ zero singular values.

Corollary 2.8.1 *The singular values of a matrix are determined uniquely.*

Corollary 2.8.2 *If $\sigma_1 > \ldots > \sigma_r > 0$, then the singular vectors $u_1, \ldots, u_r$ and $v_1, \ldots, v_r$ are determined uniquely up to a factor equal to 1 in modulus.*

Corollary 2.8.3

$$A u_i = \begin{cases} \sigma_i v_i, & 1 \leq i \leq r, \\ 0, & r+1 \leq i \leq n. \end{cases} \tag{2.8.3}$$

$$A^* v_i = \begin{cases} \sigma_i u_i, & 1 \leq i \leq r, \\ 0, & r+1 \leq i \leq m. \end{cases} \tag{2.8.4}$$

Corollary 2.8.4 $A = \sum\limits_{i=1}^{r} \sigma_i v_i u_i^*.$

Corollary 2.8.5

$$\begin{aligned}
\ker A &= \operatorname{span}\{u_{r+1}, \ldots, u_n\}, \\
\operatorname{im} A &= \operatorname{span}\{v_1, \ldots, v_r\}, \\
\ker A^* &= \operatorname{span}\{v_{r+1}, \ldots, v_m\}, \\
\operatorname{im} A^* &= \operatorname{span}\{u_1, \ldots, u_r\}
\end{aligned}$$

2.9 Unitarily invariant norms

If $\|A\| = \|QAZ\|$ for any unitary Q, Z and any matrix A (provided that the sizes accord), then the matrix norm involved is called *unitarily invariant*.

The most important unitarily invariant norms are $\|A\|_2$ and $\|A\|_F$. Here is the proof:

$$\begin{aligned}
\|QAZ\|_2 &= \sup_{x \neq 0} \frac{\|QAZx\|_2}{\|x\|_2} = \sup_{x \neq 0} \frac{\|(QAZ)Z^* x\|_2}{\|Z^* x\|_2} = \\
&= \sup_{x \neq 0} \frac{\|QAx\|_2}{\|x\|_2} = \sup_{x \neq 0} \frac{\|Ax\|_2}{\|x\|_2} = \|A\|_2;
\end{aligned}$$

$$\|QAZ\|_F^2 = \operatorname{tr}(QAZ)^*(QAZ) = \operatorname{tr} Z^* (A^* A) Z = \operatorname{tr} A^* A = \|A\|_F^2. \quad \square$$

Due to (2.8.1), for any unitarily invariant norm we have $\|A\| = \|\Sigma\|$, i.e., the unitarily invariant norm of a matrix is determined entirely by its singular values. Concerning the spectral and the Frobenius norms, we find that

$$\|A\|_2 = \sigma_1; \tag{2.9.1}$$

$$\|A\|_F = \left(\sigma_1^2 + \ldots + \sigma_r^2\right)^{1/2}. \tag{2.9.2}$$

2.10 A short way to the SVD

The norm $\|A\|_2$ appeared as an operator norm. From the compactness of the unit sphere, there exist normalized vectors x and y for which $Ax = \sigma y$, $\sigma = \|A\|_2$. Choose unitary matrices of the form $U = [x \;\; U_1]$, $V = [y \;\; V_1]$. Then

$$V^* A U = \begin{bmatrix} \sigma & w^* \\ 0 & B \end{bmatrix},$$

$$\left\|V^* A U \begin{bmatrix} \sigma \\ w \end{bmatrix}\right\|_2^2 \geq \left(\sigma^2 + w^* w\right)^2 \Rightarrow \|V^* A U\|_2^2 \geq \sigma^2 + w^* w.$$

Since $\|V^* A U\|_2 = \|A\|_2$ (the spectral norm is unitary invariant), $w = 0$. Proceed by induction.

2.11 Approximations of a lower rank

Theorem 2.11.1 *Suppose* $k < \operatorname{rank} A$, $\quad A_k \equiv \sum_{i=1}^{k} \sigma_i v_i u_i^*$. *Then*

$$\min_{\operatorname{rank} B = k} \|A - B\|_2 = \|A - A_k\|_2 = \sigma_{k+1}.$$

Proof. Since $\operatorname{rank} B = k$, we find that $\dim \ker B = n - k$. Suppose that

$$\ker B = \operatorname{span}\{x_1, \ldots, x_{n-k}\}.$$

Then there is a nonzero $z \in \operatorname{span}\{x_1, \ldots, x_{n-k}\} \cap \operatorname{span}\{u_1, \ldots, u_{k+1}\}$. (Why?) Assume that $\|z\|_2 = 1$. Then

$$\|A - B\|_2^2 \geq \|(A - B)z\|_2^2 = \|Az\|_2^2 = \sum_{i=1}^{k+1} \sigma_i \left(u_i^* z\right)^2 \geq \sigma_{k+1}^2. \quad \square$$

In particular, the inferior singular value of a nonsingular matrix is the distance (in the spectral norm) to the nearest singular one.

2.12   Smoothness and ranks

Sometimes the entries of a matrix are regarded as the values of a function at some points of a mesh. For sufficiently smooth functions, such matrices are close to low-rank matrices.

Suppose that $f(x, y)$ has an infinite number of derivatives in $y \in Y \subset \mathbb{R}$ for any $x \in X \subset \mathbb{R}$. Consider the two suites of points

$$x_1^{(n)}, \dots, x_n^{(n)} \in X, \quad y_1^{(n)}, \dots, y_n^{(n)} \in Y,$$

and assume that

$$A_n = [f(x_i^{(n)}, y_j^{(n)})], \quad 0 \leq i, j \leq n.$$

Approximate $f(x, y)$ by a truncated Taylor series at some point $z \in Y$:

$$f(x, y) \approx f_p(x, y) \equiv \sum_{k=0}^{p-1} \frac{\partial_k f(x, z)}{k!} (y - z)^k, \quad \partial_k f \equiv \left(\frac{\partial}{\partial y} \right)^k f,$$

and consider the following matrix:

$$\tilde{A}_n \equiv [f_p(x_i^{(n)}, y_j^{(n)})], \quad 0 \leq i, j \leq n.$$

Of course $A_n \approx \tilde{A}_n$. What is more, $\tilde{A}_n$ has a rank that is low compared to n:

$$\operatorname{rank} \tilde{A}_n \leq p. \quad \text{(Prove this.)}$$

It is remarkable that this inequality holds no matter how large n is.

To apply the above observation, we need know a bit more about the relation between the ranks and the approximation accuracy. To this end, the well-known remainder term estimates for the Taylor series may serve. For example, prove that if $|(\partial / \partial y)^k f| \leq M \quad \forall k$, then for any $\varepsilon > 0$,

$$\|A_n - \tilde{A}_n\|_F = O(n\varepsilon) \quad \text{while} \quad p = O(\log \varepsilon^{-1}).$$

The assumption we made can be relaxed in many ways.

Exercises

1. Prove that in the case $p \neq 2$, $p \geq 1$ the norm $\|x\|_p$ cannot be induced by any scalar product.

2. Prove that for any matrix $A \in \mathbb{C}^{m \times n}$, the subspaces $\ker A$ and $\operatorname{im} A^*$ are orthogonal and the whole of $\mathbb{C}^n$ is their direct sum.

3. Prove that $\|A\|_F = \|A\|_2$ if and only if $\operatorname{rank} A = 1$.

4. A matrix A^{1997} is normal. Must A be normal?

5. Prove that for any Hermitian matrix H, the matrix

$$Q = (I - \mathrm{i}\, H)^{-1}(I + \mathrm{i}\, H)$$

will be unitary. Is it true that an arbitrary unitary matrix can be written in this way?

6. Let $A = I + \alpha u u^*, \|u\|_2 = 1$. Find all those α for which A is unitary.

7. Suppose $A^* = I + \beta A$. Prove that if A has at least two different eigenvalues, then $|\beta| = 1$.

8. Prove that a matrix which is the product of a Hermitian matrix and a positive definite (Hermitian) matrix has all its eigenvalues real.

9. Calculate the SVD for the following $n \times n$ matrix:

$$A = \begin{bmatrix} 1 \\ 2 \\ \cdots \\ n \end{bmatrix} [1\ 1 \dots\ 1].$$

10. What is the distance between a singular matrix A and the nearest non-singular one?

11. Is it valid that $B = A^*$ holds if $(Ax, x) = (x, Bx)$ for any: (a) $x \in \mathbb{R}^n$ or (b) $x \in \mathbb{C}^n$?

12. Prove that for a square matrix A, $\lambda_{min}(A + A^*) \le 2\sigma_{min}(A)$, where $\lambda_{min}(\cdot)$ and $\sigma_{min}(\cdot)$ denote the minimal eigenvalue and the minimal singular value. Would this be still true if λ_{min} from the left-hand side were changed onto σ_{min}?

13. Prove that $\|AB\|_F \le \|A\|_2 \|B\|_F$.

14. Prove that $\|A\|_F \le \sqrt{\text{rank}(A)}\, \|A\|_2$.

15. A matrix $A = [A_{ij}]$ is composed of the blocks A_{ij}, and a matrix $B = [b_{ij}]$ is such that $b_{ij} = \|A_{ij}\|_2$, $1 \le i \le m$, $1 \le j \le n$. Prove that

$$\|A\|_2 \le \|B\|_2.$$

16. Suppose that L is the lower triangular part of a matrix $A \in \mathbb{C}^{n\times n}$. Prove that

$$\|L\|_2 \le \log_2 2n\, \|A\|_2.$$

17. Let $A \in \mathbb{C}^{n\times n}$. Prove that $\text{tr}\, A = 0$ if and only if $\|I + zA\|_F \ge \sqrt{n}$ for all $z \in \mathbb{C}$.

18. Assume that $A_{n \times n}$ is a Hermitian nonnegatively definite matrix. Prove that

$$\operatorname{tr} A \leq \sqrt{\operatorname{rank} A} \, \|A\|_F.$$

19. Let a subspace $L \in \mathbb{C}^n$ be fixed, and consider matrices P such that $P^2 = P$ and im $P = L$. Prove that the least value of the 2-norm over all such matrices P is attained at a Hermitian matrix.

Lecture 3

3.1 Perturbation theory

Suppose that we are to compute $f(x)$ at some x. We know that algorithms sometimes do not produce very accurate answers. When thinking this over, we should comprehend that not only might an algorithm be "bad" but it might be a problem itself. An important question: how far can $f(x)$ change when x goes through small perturbations?

In the simplest case, $f(x + \delta) \simeq f(x) + f'\delta$, and hence, the value $||f'||$ can serve as an "absolute" measure of the problem's sensitivity. If $f(x) \neq 0$ and $x \neq 0$, then

$$\frac{f(x+\delta) - f(x)}{||f(x)||} \simeq \left(\frac{f'(x)}{||f(x)||} ||x|| \right) \frac{\delta}{||x||}.$$

Consequently, a relative measure of the problem's sensitivity (in other words, its *condition number*) can be defined as

$$\mathrm{cond}\,(f(x)) \equiv \frac{||f'(x)||}{||f(x)||} ||x||.$$

3.2 Condition of a matrix

Let A be a nonsingular matrix and $f(A) = A^{-1}$. Then (please check it)

$$(A + \Delta)^{-1} - A^{-1} = -A^{-1}\Delta(A + \Delta)^{-1} \simeq -A^{-1}\Delta A^{-1}$$

$$\Rightarrow \quad \frac{||(A + \Delta)^{-1} - A^{-1}||}{||A^{-1}||} \lesssim (||A^{-1}||\,||A||) \frac{||\Delta||}{||A||}.$$

The quantity

$$\mathrm{cond}(A) \equiv ||A^{-1}||\,||A||$$

is called the *condition number* of A. It depends on the norm. In the case of the p-norm, we write cond_p. Usually, cond_2 is called the spectral condition number.

For singular matrices, it is natural to set cond $= \infty$. Usually, the condition number of a problem is inversely proportional to the distance between it and the set of singular (in a proper sense) problems.[1] If we consider the inversion of matrices, then the set of singular problems is that of singular matrices. We already know that

$$\min_{\det S=0} ||A - S||_2 = \sigma_{\min} \quad \text{(the minimal singular value)}$$

and, simultaneously (prove this), $||A^{-1}||_2 = 1 \, / \, \sigma_{\min}$. Hence,

$$\text{cond } (A) = \frac{||A||_2}{\min\limits_{\det S=0} ||A - S||_2}.$$

Why does $(A + \Delta)^{-1} \approx A^{-1}$ for small Δ ? This follows from standard continuity considerations. In matrix analysis, however, there is a simple and useful framework for such cases.

3.3   Convergent matrices and series

A series $\sum_{k=0}^{\infty} A_k$, where $A_k \in \mathbb{C}^{n \times n}$, is called convergent if the sequence of its partial sums $S_N \equiv \sum_{k=0}^{N} A_k$ is such. This follows from the convergence of the numeric series $\sum_{k=0}^{\infty} ||A_k||$ (prove this).

A series of the form $\sum_{k=0}^{\infty} F^k$ is called the *Neumann series*. Obviously, it converges whenever $||F|| < 1$ (prove this). It is less obvious that it converges whenever the eigenvalues of F in modulus are less than 1.

The maximal in modulus eigenvalue of a matrix (say, F) is called its *spectral radius*. The notation: $\rho(F)$. If $\rho(F) < 1$, then the matrix F is called *convergent*.

Lemma 3.3.1 *The Neumann series for a matrix $F \in \mathbb{C}^{n \times n}$ converges if and only if the matrix F is convergent.*

Sufficiency. From the Schur theorem, for some unitary matrix P, the matrix $T = [t_{ij}] = P^{-1}FP$ is upper triangular. We show that the Neumann series is convergent for some matrix which is similar to F. (Would you explain why this is equivalent to its convergence for F?)

Set $D_\varepsilon = \text{diag}(1, \varepsilon, \ldots, \varepsilon^{n-1})$. Then $\{D_\varepsilon^{-1} T D_\varepsilon\}_{ij} = \varepsilon^{j-i} t_{ij}$ for $i \leq j$. The diagonal elements of this matrix in modulus are less then 1 $\Rightarrow$ for sufficiently small ε $||D_\varepsilon^{-1} T D_\varepsilon||_1 < 1 \Rightarrow$ the Neumann series for the matrix $D_\varepsilon^{-1} T D_\varepsilon$ is convergent.

Necessity. Assume that $Fx = \lambda x$, $x \neq 0$, and $|\lambda| > 1$. Then $|\lambda|^k \leq ||F^k||_2$ (why?) $\Rightarrow ||F^k||_2 \to 0 \Rightarrow$ the Neumann series for F is convergent. $\square$

---

[1] For more detail, see   J. W. Demmel. On condition numbers and the distance to the nearest ill-posed problem. *Numer. Math.* 51, 251–289 (1987).

3.4 The simplest iteration method

To solve a linear algebraic system $Ax = b$ with a nonsingular coefficient matrix, we rewrite it as $x = Fx + \alpha\, b$, where $F = I - \alpha\, A$, $\alpha \neq 0$, and consider the following iterative method:

$$x_0 \quad \text{is an arbitrary initial vector;}$$
$$x_k \;=\; Fx_{k-1} + \alpha b \quad \text{for} \quad k = 1, 2, \ldots\ .$$

This is the so-called *simplest iteration method* (sometimes it is referred to as Richardson's method).

It is easy to deduce (with Lemma 3.3.1) that $x_k \to x$ for any initial vector x_0 if and only if the matrix F is convergent.

A good idea is to find a splitting $A = M - N$ for which the matrix $M^{-1}N$ would be convergent while M is *easily invertible* (that is, there is an efficient way to solve systems with the coefficient matrix M). Then

$$x_k = M^{-1}(Nx_{k-1} + b) \;\to\; x.$$

3.5 Inverses and series

Lemma 3.5.1 *If $\|F\| < 1$, then the matrix $A = I - F$ is nonsingular, and the inverse enjoys the following properties:*

$$\text{(a)} \quad (I - F)^{-1} = \sum_{k=0}^{\infty} F^k; \qquad \text{(b)} \quad \|(I - F)^{-1}\| \leq \frac{\|I\|}{1 - \|F\|}.$$

Proof. It is easy to verify that

$$(I - F)\left(\sum_{k=0}^{N} F^k\right) = I - F^{N+1} \longrightarrow I.$$

To prove (b), let us write

$$\left\|\sum_{k=0}^{N} F^k\right\| \leq \|I\| \sum_{k=0}^{N} \|F\|^k \;\leq\; \frac{\|I\|}{1 - \|F\|}. \quad \square$$

Corollary 3.5.1 *If A is a nonsingular matrix and E is a perturbation such that $\|A^{-1}E\| < 1$, then*

(a) *the matrix $A + E$ is nonsingular and*

$$(A + E)^{-1} = \sum_{k=0}^{\infty} \left(-A^{-1}E\right)^k A^{-1} \;=\; A^{-1} \sum_{k=0}^{\infty} \left(-EA^{-1}\right)^k;$$

(b) $\quad \dfrac{\|(A + E)^{-1} - A^{-1}\|}{\|A^{-1}\|} \;\leq\; \dfrac{\|A^{-1}\|\|E\|}{1 - \|A^{-1}E\|}.$

3.6   Condition of a linear system

Consider a system $Ax = f$, $f \neq 0$, with a nonsingular matrix A and a perturbed system $(A + \Delta A)\tilde{x} = f + \Delta f$. How far can $\tilde{x}$ differ from x? Assume that $\|A^{-1}\Delta A\| < 1$. Then

$$
\begin{aligned}
\tilde{x} - x &= (A + \Delta A)^{-1}(f + \Delta f) - A^{-1}f \\
&= \left[(A + \Delta A)^{-1} - A^{-1}\right] f + (A + \Delta A)^{-1}\Delta f \\
&= \left[\sum_{k=1}^{\infty}(-A^{-1}\Delta A)^k\right] (A^{-1}f) + \left[\sum_{k=0}^{\infty}(-A^{-1}\Delta A)^k\right] A^{-1}\Delta f
\end{aligned}
$$

$$
\Rightarrow \qquad \frac{\|\tilde{x} - x\|}{\|x\|} \leq \frac{\|A^{-1}\|\|A\|}{1 - \|A^{-1}\Delta A\|}\left(\frac{\|\Delta A\|}{\|A\|} + \frac{\|\Delta f\|}{\|f\|}\right). \qquad (3.6.1)
$$

The value cond $A = \|A^{-1}\|\|A\|$ (the condition number of A) is a characteristic of the sensitivity of the solution x to small perturbations of the right-hand side. Matrices with "too large" and "not too large" condition numbers are said to be ill-conditioned and well-conditioned matrices.

3.7   Consistency of matrix and right-hand side

The bound (3.6.1) cannot be improved on the whole set of matrices and perturbations. However, the ill condition of a matrix is not quite the same as "the ill condition of a linear system".

Using the singular value decomposition,

$$
A = \sum_{k=1}^{n} \sigma_k v_k u_k^* \quad \Rightarrow \quad A^{-1} = \sum_{k=1}^{n} \sigma_k^{-1} u_k v_k^* \quad \Rightarrow \quad x = \sum_{k=1}^{n} \frac{v_k^* f}{\sigma_k} u_k.
$$

Suppose that A is fixed and $\Delta f = \xi_1 v_1 + \cdots + \xi_n v_n$, $\Delta x = \eta_1 u_1 + \cdots + \eta_k u_k$. Clearly,

$$
\sigma_1^2 \eta_1^2 + \cdots + \sigma_n^2 \eta_n^2 = \xi_1^2 + \cdots + \xi_n^2.
$$

This implies that if Δf belongs to the ball of the radius ε in the coordinates $\{\xi_i\}$, then Δx belongs to the ellipsoid in the coordinates $\{\eta_i\}$:

$$
\xi_1^2 + \ldots + \xi_n^2 \leq \varepsilon^2 \quad \Leftrightarrow \quad \frac{\eta_1^2}{1/\sigma_1^2} + \ldots + \frac{\eta_n^2}{1/\sigma_n^2} \leq \varepsilon^2.
$$

Thus we see that $\|\Delta x\|$ depends dramatically on the direction of perturbations.

If the right-hand side f has nonzero components along the inferior singular vector $v_{r+1}, \ldots, v_n$ (in this case, one says that a matrix and a right-hand side are *consistent*) and the perturbations are zero along the same directions, then the bound (3.6.1), obviously, becomes better:

$$\frac{\|\Delta x\|_2}{\|x\|_2} \leq \frac{\sigma_1}{\sigma_r} \frac{\|\Delta f\|_2}{\|f\|_2}.$$

3.8 Eigenvalue perturbations

Let $\lambda(A)$ denote the spectrum of a matrix A.

The Bauer–Fike theorem. *If $\mu \in \lambda(A + F)$ but $\mu \notin \lambda(A)$, then*

$$\frac{1}{\|(A - \mu I)^{-1}\|_2} \leq \|F\|_2.$$

Proof. The matrix $(A + F) - \mu I = (A - \mu I) + F$ is singular $\Rightarrow$ the matrix $I + (A - \mu I)^{-1}F$ is also singular $\Rightarrow$ $\|(A - \mu I)^{-1}F\|_2 \geq 1$. $\square$

Theorem 3.8.1 *Assume that A is diagonalizable:*

$$P^{-1}AP = \operatorname{diag}(\lambda_1, \ldots, \lambda_n) \equiv \Lambda. \tag{3.8.1}$$

Then, if $\mu \in \lambda(A + F)$, then

$$\min_{1 \leq i \leq n} |\mu - \lambda_i| \leq \|P^{-1}\|_2 \|P\|_2 \|F\|_2. \tag{3.8.2}$$

Proof. The inequality is trivial if $\mu \in \lambda(A)$. Otherwise, if $\mu \notin \lambda(A)$, then $\mu \notin \lambda(\Lambda)$ and $\mu \in \lambda(\Lambda + P^{-1}FP)$, and it remains to apply the Bauer–Fike theorem. $\square$

Thus, the sensitivity of the spectrum to small perturbations is characterized by the condition of the eigenvector matrix P (the columns are the eigenvectors of A).

Theorem 3.8.2 *Let $P^{-1}AP = J$ be the Jordan matrix for A and $\mu \in \lambda(A + F)$. Then there exists $\lambda \in \lambda(A)$ such that*

$$\frac{|\mu - \lambda|^m}{1 + |\mu - \lambda| + \ldots + |\mu - \lambda|^{m-1}} \leq \|P^{-1}\|_2 \|P\|_2 \|F\|_2,$$

where m is the maximal order of the Jordan blocks corresponding to λ.

Proof. Again use the Bauer–Fike theorem: if $\mu \notin \lambda(A)$, then

$$\frac{1}{\|(J - \mu I)^{-1}\|_2} \leq \|P^{-1}FP\|_2.$$

Suppose J consists of the Jordan blocks $J_1, \ldots, J_k$. Then

$$\frac{1}{\|(J - \mu I)^{-1}\|_2} \geq \frac{1}{\max\limits_{1 \leq i \leq k} \|(J_i - \mu I)^{-1}\|_2}.$$

Write $J_i = \lambda I + N_i$, and assume that this Jordan block is of order m. Then $N_i^m = 0$ and, moreover, $||N_i||_2 = 1 \quad \Rightarrow$

$$||(J_i - \mu I)^{-1}||_2 = ||((\lambda - \mu) I + N_i)^{-1}||_2 \leq ||(I + (\lambda - \mu)^{-1} N_i)^{-1}||_2 |(\lambda - \mu)^{-1}|$$

$$\leq \left(1 + |(\lambda - \mu)^{-1}| + \ldots + |(\lambda - \mu)^{-1}|^{m-1}\right) |(\lambda - \mu)^{-1}|. \quad \Box$$

Thus, if a matrix with the maximal order of its Jordan blocks equal to m is perturbed by quantities of an ε order of magnitude, then any eigenvalue of the perturbed matrix differs from some eigenvalue of the original matrix by the value of the $|\varepsilon|^{\frac{1}{m}}$ order of magnitude.

Is it true that, for sufficiently small perturbations, the eigenvalues of A and $A + F$ can be divided into pairs of close ones? By way of answering, we can say that it is so since the roots of a polynomial are continuous functions of its coefficients. The latter fact itself deserves a special discussion.

3.9 Continuity of polynomial roots

Theorem 3.9.1 *Consider a parametrized batch of polynomials*

$$p(x, t) = x^n + a_1(t)x^{n-1} + \cdots + a_n(t),$$

where $a_1(t), \ldots, a_n(t) \in C[\alpha, \beta]$. *Then there exist functions*

$$x_1(t), \ldots, x_n(t) \in C[\alpha, \beta]$$

such that

$$p(x_i(t), t) = 0 \quad for \quad \alpha \leq t \leq \beta, \qquad i = 1, \ldots, n.$$

To begin the proof, note that it is sufficient to establish the existence of any *single* continuous function $x_n(t)$ such that $p(x_n(t), t) = 0$ for $\alpha \leq t \leq \beta$. Should this be done, we write

$$p(x, t) = (x - x_n(t)) q(x, t),$$

where $q(x, t) = x^{n-1} + b_1(t)x^{n-2} + \cdots + b_{n-1}(t)$. On the strength of the familiar algorithm for dividing polynomials, $b_1(t), \ldots, b_{n-1}(t) \in C[\alpha, \beta]$. So we may prove it by induction.

Now, we shall prove the existence of one continuous root. To do this, remember the proof of the existence of a solution to a differential equation $\frac{dy}{dt} = f(t, y)$ in the case of a continuous f by the Euler piecewise linear functions and the Arzela-Ascoli theorem.

A sequence of functions $y_m(t)$ is called *uniformly continuous* for $t \in [\alpha, \beta]$ if $\forall \varepsilon > 0 \exists \delta > 0 : |t_1 - t_2| \leq \delta \Rightarrow |y_m(t_1) - y_m(t_2)| \leq \varepsilon \forall m$. A sequence of functions $y_m(t)$ is called *uniformly bounded* for $t \in [\alpha, \beta]$ if $\exists c > 0 : |y_m(t)| \leq c \forall m, \forall t \in [\alpha, \beta]$.

Theorem 3.9.2 (Arzela-Ascoli) *For any sequence of uniformly continuous and uniformly bounded functions on $[\alpha, \beta]$ there exists a subsequence which converges uniformly on $[\alpha, \beta]$.*

Proof. Enumerate, in any way, all the rational points on $[\alpha, \beta]$: $t_1, t_2, \ldots$. From the original sequence $y_m(t)$, we choose, first, a subsequence $y_{1,m}(t)$ convergent at some point t_1; from the latter, we choose a subsequence $y_{2,m}(t)$ convergent at some point t_2, and so on. In the end, we will have the "nested" subsequences $y_{1,m}(t), \ldots, y_{k,m}(t), \ldots$ such that $y_{k,m}(t)$ converges at $t = t_1, \ldots, t_k$ (and $y_{k+1,m}(t)$ is a subsequence of $y_{k,m}(t)$). Consider the "diagonal" sequence $y_{m,m}(t)$. Take any $\varepsilon > 0$, and choose $\delta > 0$ determined by the uniform continuity property. For an arbitrary point $t \in [\alpha, \beta]$, there exists t_i such that $|t - t_i| \leq \delta$. For sufficiently large m, k, we obtain

$$
\begin{aligned}
|y_{mm}(t) - y_{kk}(t)| &\leq |y_{mm}(t) - y_{mm}(t_i)| + |y_{mm}(t_i) - y_{kk}(t_i)| + \\
&\quad + |y_{kk}(t_i) - y_{kk}(t)| \\
&\leq 2\varepsilon + |y_{mm}(t_i) - y_{kk}(t_i)| \leq 3\varepsilon,
\end{aligned}
$$

which means that the $y_{m,m}(t)$ is the Cauchy sequence. $\square$

Proof of Theorem 3.9.1. Build up on $[\alpha, \beta]$ a sequence of uniform grids

$$
\alpha = t_{0m} < t_{1m} < \ldots < t_{mm} = \beta; \quad t_{i+1,m} - t_{im} = \frac{\beta - \alpha}{m}.
$$

Let $y_m(t)$ be a piecewise linear function with breaks at $t_{0m}, t_{1m}, \ldots, t_{mm}$. Define the values at the nodes as follows.

Take a root z_0 of the polynomial $p(x, \alpha)$, and, for all m, set

$$
y_m(t_{0m}) = z_{0m} \equiv z_0.
$$

Further, let z_{1m} be any of those roots of the polynomial $p(x, t_{1m})$ nearest to z_{0m}, and, by induction, let $z_{i+1,m}$ be any of the roots of the polynomial $p(x, t_{i+1,m})$ nearest to z_{im}. Set

$$
y_m(t_{im}) = z_{im}, \qquad i = 1, \ldots, m.
$$

The uniform boundedness of the piecewise linear functions $y_m(t)$ is evident. The uniform continuity emanates from the inequality

$$
|z_{i+1,m} - z_{im}| \leq |p(z_{im}, t_{i+1,m})|^{\frac{1}{n}} = |p(z_{im}, t_{i+1,m}) - p(z_{im}, t_{im})|^{\frac{1}{n}}
$$

$$
\leq R \left(\max_{\substack{\alpha \leq t_1, t_2 \leq \beta \\ |t_1 - t_2| \leq \frac{\beta - \alpha}{m}}} \sum_{j=1}^{n} |a_j(t_1) - a_j(t_2)| \right)^{\frac{1}{n}},
$$

where $R \geq 1$ is the radius (not necessary minimal) of a circle encompassing all the roots of all the polynomials $p(x, t)$ for $\alpha \leq t \leq \beta$.

Using the Arzela-Ascoli theorem we find a uniformly convergent subsequence. Take into account that the limit of a uniformly convergent sequence of continuous functions on $[\alpha, \beta]$ must be a continuous function. The only thing left to check is that the limit function $y(t)$ satisfies $p(y(t), t) = 0$ for $\alpha \le t \le \beta$. That will do the proof. $\quad\square$

Exercises

1. That the roots depend continuously on a polynomial coefficients no longer means that they cannot vary dramatically after some small perturbations. Here is Wilkinson's example:

$$p(x; \varepsilon) = (x - 1)(x - 2) \ldots (x - 20) + \varepsilon\, x^{19} = \prod_{i=1}^{20} (x - x_i(\varepsilon)),$$

   where $x_i(\varepsilon)$ are continuous functions such that $x_i(0) = i$. For small ε, we observe that $x_1(\varepsilon) \approx 1$, but $x_{20}(\varepsilon)$ goes far from 20. To explain this, compare the values of derivatives of the functions $x_1(\varepsilon)$ and $x_{20}(\varepsilon)$ at the point $\varepsilon = 0$.

2. Does $|\det A| = 1$ mean that A is well conditioned? Does $|\det A| \ll 1$ mean that A is ill conditioned?

3. Calculate $\text{cond}_\infty(A) = \|A^{-1}\|_\infty \|A\|_\infty$ for the bidiagonal matrix

$$A(\varepsilon) = \begin{bmatrix} 1 & 2 & & & 0 \\ & 1 & 2 & & \\ & & \ddots & \ddots & \\ & & & 1 & 2 \\ 0 & & & & 1 \end{bmatrix}_{n \times n}.$$

4. Let $\rho(A)$ be the spectral radius of a matrix A. Prove that any operator norm $\|\cdot\|$,

$$\rho(A) = \lim_{n \to \infty} \|A^n\|^{\frac{1}{n}}.$$

5. Denote by $\lambda_{\min}(\cdot)$ the minimal in modulus eigenvalue and by $\sigma_{\min}(\cdot)$ the minimal singular value of a matrix. Prove that for any $A \in \mathbb{C}^{n \times n}$,

$$|\lambda_{\min}(A)| = \lim_{n \to \infty} (\sigma_{\min}(A^n))^{\frac{1}{n}}.$$

6. Suppose A is an arbitrary nonsingular matrix. Is it always possible to pick up some α so that the matrix $I - \alpha A$ would be convergent?

7. Suppose A is a Hermitian positive definite matrix. Prove that, for all α sufficiently small, the matrix $I - \alpha A$ is convergent.

8. Suppose that $A = \alpha I - N$, where N is a square matrix with nonnegative elements and $\alpha > |\lambda_i(N)|$ for all the eigenvalues of N. Prove that the matrix A is nonsingular and all elements of A^{-1} are nonnegative.

9. Is it true that, for any matrix A, there is a splitting $A = M - N$ with the convergent matrix $M^{-1}N$?

10. Let M be nonsingular, and let $M^*M - N^*N$ be nonnegative-definite. Prove that
$$\rho(M^{-1}N) \leq 1$$
($\rho(\cdot)$ is the spectral radius of a matrix).

11. Consider a polynomial $p(x) = x^n + a_1 x^{n-1} + \cdots + a_n$ and the so-called Frobenius matrices of the following form:

$$\Phi = \begin{bmatrix} 0 & 0 & 0 & \cdots & 0 & a_n \\ -1 & 0 & 0 & \cdots & 0 & a_{n-1} \\ 0 & -1 & 0 & \cdots & 0 & a_{n-2} \\ \cdots & \cdots & \cdots & \cdots & 0 & a_1 \\ 0 & 0 & 0 & \cdots & -1 & 0 \end{bmatrix}$$

and

$$\Psi = \begin{bmatrix} 0 & -1 & 0 & \cdots & 0 & 0 \\ a_1 & 0 & -1 & \cdots & 0 & 0 \\ a_2 & 0 & 0 & \cdots & 0 & 0 \\ \cdots & \cdots & \cdots & \cdots & 0 & -1 \\ a_n & 0 & 0 & \cdots & 0 & 0 \end{bmatrix}.$$

Prove (easiest by induction) that $p(x)$ is the characteristic polynomial both for Φ and Ψ. Is it possible to transform Theorem 3.8.1 into a theorem about the perturbation of the polynomial roots?

12. Find the eigenvalues of the perturbed Jordan block

$$J(\varepsilon) = \begin{bmatrix} \lambda & 1 & & & 0 \\ & \lambda & 1 & & \\ & & \ddots & \ddots & \\ & & & \lambda & 1 \\ \varepsilon & & & & \lambda \end{bmatrix}_{n \times n}.$$

13. Assume that the eigenvalues of a real symmetric matrix A are pairwise distinct. Prove that, for all perturbations F sufficiently small in norm, the eigenvalues of the perturbed matrix $A + F$ are real.

Lecture 4

4.1 Diagonal dominance

A matrix $A \in \mathbb{C}^{n \times n}$ is *row-wise diagonally dominant* if

$$|a_{ii}| > r_i \equiv \sum_{\substack{j=1 \\ j \neq i}}^{n} |a_{ij}|, \quad i = 1, \ldots, n, \tag{4.1.1}$$

and *column-wise diagonally dominant* if

$$|a_{jj}| > c_j \equiv \sum_{\substack{i=1 \\ i \neq j}}^{n} |a_{ij}|, \quad j = 1, \ldots, n. \tag{4.1.2}$$

Theorem 4.1.1 (Levy–Desplanques) *If a matrix is row-wise or column-wise diagonally dominant, then it is nonsingular.*

Proof. Adopt the following notation:

$$\mathrm{diag}\,(A) \equiv \mathrm{diag}\,(a_{11}, \ldots, a_{nn}), \qquad \mathrm{off}\,(A) \equiv A - \mathrm{diag}\,(A).$$

Then the inequalities (4.1.1) imply that $\|[\mathrm{diag}\,(A)]^{-1}\,\mathrm{off}\,(A)\|_{\infty} < 1$ and, thence, a matrix $A = \mathrm{diag}\,(A) + \mathrm{off}\,(A)$ is nonsingular. The case (4.1.2) reduces to the case (4.1.1) by transition to A^{T}. $\square$

4.2 Gerschgorin disks

Theorem 4.2.1 (Gerschgorin) *For a matrix $A \in \mathbb{C}^{n \times n}$, consider the disks*

$$R_i \equiv \{z \in \mathbb{C} : |a_{ii} - z| \leq r_i\},$$
$$C_i \equiv \{z \in \mathbb{C} : |a_{ii} - z| \leq c_i\},$$

where r_i and c_i are defined by (4.1.1) and (4.1.2), respectively. Then, if $\lambda \in \lambda(A)$, then

$$\lambda \in \left(\bigcup_{i=1}^{n} R_i \right) \cap \left(\bigcup_{i=1}^{n} C_i. \right)$$

Proof. If $\lambda \notin \bigcup R_i$, then the matrix $A - \lambda I$ is row diagonally dominant and, hence (by the Levy–Desplanques theorem), nonsingular $\Rightarrow \lambda \notin \lambda(A)$. □

Theorem 4.2.2 *Let G be the union of m Gerschgorin discs. If G is isolated from the other disks, then G contains exactly m eigenvalues.*

Proof. Set $A(t) = \text{diag}(A) + t \text{off}(A), 0 \leq t \leq 1$, and denote by $G(t)$ the union of the Gerschgorin discs with the same centers as those in G, and by $\bar{G}(t)$ the union of the remainder discs; $\bar{G} \equiv \bar{G}(1)$. Obviously, $G(t) \subset G$ and $\bar{G}(t) \subset \bar{G}$.

Since $G \cap \bar{G} = \emptyset$, $G(t) \cap \bar{G}(t) = \emptyset$ for all $0 \leq t \leq 1$. According to the theorem on the continuous dependence of the polynomial roots on their coefficients, there exist continuous functions $\lambda_1(t), \ldots, \lambda_m(t)$ such that

(a) $\{\lambda_1(0), \ldots, \lambda_m(0)\} = G(0)$;

(b) $\lambda_1(t), \ldots, \lambda_m(t) \in \lambda(A(t))$.

Let $t_i \equiv \max\{t : \lambda_i(t) \in G(t)\}$. If $t_i < 1$, , then, for all $t > t_i$, we have $\lambda_i(t) \in \bar{G}(t)$ (it follows from Theorem 4.2.1).

Thus we conclude that $G(t_i) \cap \bar{G}(t_i) \neq \emptyset$, which is impossible. Therefore, $t_i = 1$ for $i = 1, \ldots, m$. □

Corollary 4.2.1 *If the Gerschgorin discs are pairwise disjoint, then each one captures exactly one eigenvalue.*

4.3   Small perturbations of eigenvalues and vectors

Assume that a matrix A has only simple (pairwise distinct) eigenvalues, and a perturbed matrix is of the form

$$A(\varepsilon) = A + A_1 \varepsilon + O\left(\varepsilon^2\right).$$

Let P be the eigenvector matrix for A; then $\Lambda \equiv P^{-1}AP$ is the diagonal matrix of the eigenvalues of A. Set

$$\Omega(\varepsilon) \equiv P^{-1}A(\varepsilon)P = \Lambda + \Omega_1 \varepsilon + O\left(\varepsilon^2\right), \qquad \Omega_1 = P^{-1}A_1 P.$$

The eigenvalues of $\Omega(\varepsilon)$ are the same as those of $A(\varepsilon)$. By the Gerschgorin theorems, for all ε small enough, these eigenvalues are simple (prove this).

We may write the diagonal matrix of the eigenvalues of $\Omega(\varepsilon)$ and the corresponding eigenvector matrix in the following form:

$$\Lambda(\varepsilon) = \Lambda + \Lambda_1 \varepsilon + \hat{\Lambda}(\varepsilon), \qquad \Lambda(0) = \Lambda;$$

$$Z(\varepsilon) = I + Z_1 \varepsilon + \hat{Z}(\varepsilon), \qquad Z(0) = I.$$

In fact, these equations merely define $\hat{\Lambda}(\varepsilon)$ and $\hat{Z}(\varepsilon)$. Since $Z(\varepsilon)$ is not quite unique (it is such only up to normalizations of the columns), we can require that

$$\text{diag } Z(\varepsilon) = I, \quad \text{diag } Z_1 = 0 \quad \Rightarrow \quad \text{diag } \hat{Z}(\varepsilon) = 0.$$

Consider the key equation

$$\left(\Lambda + \Omega_1 \varepsilon + O\left(\varepsilon^2\right)\right)\left(I + Z_1 \varepsilon + \hat{Z}\right) = \left(I + Z_1 \varepsilon + \hat{Z}\right)\left(\Lambda + \Lambda_1 \varepsilon + \hat{\Lambda}\right). \quad (*)$$

First of all, rewrite $(*)$ as follows:

$$(\Lambda Z_1 - Z_1 \Lambda + \Omega_1 - \Lambda_1)\varepsilon + \ldots = 0,$$

and determine Λ_1 and Z_1 from the equation

$$\Lambda Z_1 - Z_1 \Lambda = \Lambda_1 - \Omega_1.$$

Obviously (just look at explicit expressions for the elements of Z_1),

$$\Lambda_1 = \text{diag } \Omega_1; \qquad \Lambda Z_1 - Z_1 \Lambda = -\text{off } \Omega_1. \qquad (**)$$

Now, let Λ_1 and Z_1 satisfy $(**)$. Consider $(*)$ as an equation for $\hat{\Lambda}$ and $\hat{Z}$. Then

$$\text{(a)} \quad \hat{\Lambda} = O\left(\varepsilon^2\right), \quad \text{(b)} \quad \hat{Z} = O\left(\varepsilon^2\right).$$

First, prove (a). To do this, note that

$$\left(\Lambda + \Omega_1 \varepsilon + O\left(\varepsilon^2\right)\right)(I + Z_1 \varepsilon) - (I + Z_1 \varepsilon)(\Lambda + \Lambda_1 \varepsilon) = O\left(\varepsilon^2\right). \qquad (***)$$

For small ε, $\left\|(I + Z_1 \varepsilon)^{-1}\right\| = O(1)$. On the strength of the Gerschgorin theorems, the eigenvalues of the matrix $\Lambda + \Omega_1 \varepsilon + O\left(\varepsilon^2\right)$ are the diagonal elements of $\Lambda_0 + \Lambda_1 \varepsilon$ within the $O\left(\varepsilon^2\right)$ accuracy. To show that (b) follows from (a), observe that $\hat{Z}$ can be regarded as a variation of the solution to the equation $(***)$ whose coefficients go through small perturbations. Thus, we obtain the following.

Theorem 4.3.1 *Assume that $P^{-1}AP = \Lambda$ is a diagonal matrix with pairwise distinct eigenvalues of A. Then, for all sufficiently small ε, the matrix $A(\varepsilon) = A + A_1 \varepsilon + O\left(\varepsilon^2\right)$ is diagonalizable:*

$$P^{-1}(\varepsilon)A(\varepsilon)P(\varepsilon) = \Lambda(\varepsilon),$$

and with this

$$\Lambda(\varepsilon) = \Lambda + \Lambda_1 \varepsilon + O\left(\varepsilon^2\right), \qquad P(\varepsilon) = P\left(I + Z_1 \varepsilon + O\left(\varepsilon^2\right)\right),$$

where

$$\Lambda_1 = \operatorname{diag}\left(P^{-1} A_1 P\right),$$

while Z_1 is such that

$$\operatorname{diag} Z_1 = 0, \qquad \Lambda Z_1 - Z_1 \Lambda = -\operatorname{off}\left(P^{-1} A_1 P\right).$$

Corollary 4.3.1 *The eigenvalues $\lambda_i(\varepsilon)$ of $A(\varepsilon)$ are of the form*

$$\lambda_i(\varepsilon) = \lambda_i + q_i^T A_1 p_i \varepsilon + O\left(\varepsilon^2\right),$$

where q_i^T are the rows of the matrix P^{-1}.

4.4   Condition of a simple eigenvalue

Assume that $\|A_1\|_2 = 1$. Then (allow for $q_i^T p_i = 1$)

$$\|q_i^T A_1 p_i\|_2 \;\leq\; \frac{\|q_i^T\|_2 \|p_i^T\|_2}{|q_i^T p_i|}.$$

The quantity

$$s(\lambda_i) \equiv \frac{\|q_i^T\|_2 \|p_i^T\|_2}{|q_i^T p_i|}$$

is called the *eigenvalue condition number* for λ_i. (The vectors p_i and q_i are the left $(Ap_i = \lambda_i p_i)$ and right $(q_i^T A = \lambda_i q_i^T)$ eigenvectors of A, respectively; the eigenvalue condition number does not depend on how the vectors p_i are q_i normalized).

The condition number is correctly defined for a simple eigenvalue even in the case of a nondiagonalizable matrix. It is significant that Corollary 4.3.1 is still valid for any simple eigenvalue provided that $q_i^T p_i = 1$. (This can be proved by transition to a slightly perturbed but ever diagonalizable matrix having the same vectors p_i and q_i for the simple eigenvalue λ_i.)

The condition of a simple eigenvalue of a matrix is related to the distance from this matrix to those for which this eigenvalue becomes multiple.

Theorem 4.4.1 (Wilkinson) *Suppose that A has a simple eigenvalue λ_i with the condition number $s(\lambda_i)$. Then there is a matrix $A + E$ for which λ_i is a multiple eigenvalue, and, what is more,*

$$\|E\|_2 \leq \frac{\|A\|_2}{\sqrt{s^2(\lambda_i) - 1}}. \tag{4.4.1}$$

Proof. Without loss of generality, we can regard A as taking on the Schur form:

$$A = \begin{bmatrix} \lambda_i & z^T \\ 0 & B \end{bmatrix}.$$

The left and right eigenvectors for λ_i are of the form

$$p = [1\ 0\ \ldots\ 0]^T \quad \text{and} \quad q^T = \begin{bmatrix} 1\ v^T \end{bmatrix} \ \Rightarrow\ s(\lambda_i) = \left(\|v\|_2^2 + 1 \right)^{1/2}.$$

Evidently, $v^T B + z = \lambda_i v^T \ \Rightarrow\ v^T(B - \lambda_i I) = z^T \ \Rightarrow\ $ the matrix $\tilde{B} \equiv B + \frac{vz^T}{\|v\|_2^2}$ has λ_i as its eigenvalue. Thus,

$$E = \begin{bmatrix} 0 & 0 \\ 0 & \frac{vz^T}{\|v\|_2^2} \end{bmatrix} \ \Rightarrow\ \|E\|_2 \leq \frac{\|z^T\|_2}{\|v\|_2} \leq \frac{\|A\|_2}{\|v\|_2} = \frac{\|A\|_2}{\sqrt{s^2(\lambda_i) - 1}}.$$

$\square$

4.5 Analytic perturbations

Assume that a series $A(\varepsilon) = \sum_{k=0}^{\infty} A_k \varepsilon^k$ converges for all $|\varepsilon| < \varepsilon_0$. In this case, if all the eigenvalues of A_0 are simple, then $A(\varepsilon)$ is diagonalizable by means of $P(\varepsilon)$:

$$P^{-1}(\varepsilon) A(\varepsilon) P(\varepsilon) = \Lambda(\varepsilon), \tag{4.5.1}$$

where

$$\Lambda(\varepsilon) = \sum_{k=0}^{\infty} \Lambda_k \varepsilon^k, \qquad P(\varepsilon) = \sum_{k=0}^{\infty} P_k \varepsilon^k. \tag{4.5.2}$$

Existence and convergence of the series $\Lambda(\varepsilon)$ for all small ε is a consequence of the analytic version of the implicit function theorem.

Matrices Λ_k and P_k are easy to determine. Set $Z_k \equiv P_0^{-1} P_k$ and $\Omega_k \equiv P_0^{-1} A_k P_0$. Then

$$(\Lambda_0 + \Omega_1 \varepsilon + \ldots)(I + Z_1 \varepsilon + \ldots) = (I + Z_1 \varepsilon + \ldots)(\Lambda_0 + \Lambda_1 \varepsilon + \ldots).$$

By equating the coefficients at ε^k, we find

$$\Lambda_0 Z_k - Z_k \Lambda_0 = \Lambda_k - \Phi_k, \tag{4.5.3}$$

where

$$\Phi_k = \sum_{i=1}^{k-1} (\Omega_i Z_{k-i} - Z_{k-i} \Omega_i) + \Omega_k. \tag{4.5.4}$$

This implies that

$$\Lambda_k = \text{diag } \Phi_k, \qquad \Lambda_0 Z_k - Z_k \Lambda_0 = -\text{off } \Phi_k. \tag{4.5.5}$$

If Λ_i and Z_i are known for $i \leq k - 1$, we are able to get Λ_k and Z_k from (4.5.5).

Consider the operator $\mathcal{A} : Z \longmapsto \Lambda_0 Z - Z \Lambda_0$ acting on the space of the matrices Z such that diag $Z = 0$. Using the simplicity of the eigenvalues, we conclude that the operator $\mathcal{A}$ is invertible. Then, for all small ε, the operator $\mathcal{A} + \varepsilon \mathcal{B}$ is also invertible, where

$$\mathcal{B} : Z \longmapsto \Omega_1(\varepsilon) Z - Z \Lambda_1(\varepsilon),$$
$$\Omega_1(\varepsilon) = \sum_{k=0}^{\infty} \Omega_{k+1} \varepsilon^k, \quad \Lambda_1(\varepsilon) = \sum_{k=0}^{\infty} \Lambda_{k+1} \varepsilon^k.$$

Set

$$Z_1(\varepsilon) = \sum_{k=0}^{\infty} Z_{k+1} \varepsilon^k.$$

Then

$$[\mathcal{A} + \varepsilon \mathcal{B}] Z_1(\varepsilon) = \Lambda_1(\varepsilon) - \Omega_1(\varepsilon).$$

We see from above that $Z_1(\varepsilon)$ is expressed by a convergent power series (should $\Lambda_1(\varepsilon)$ also be expressed like this).

If A_0 has multiple eigenvalues, then the eigenvalues and vectors can be expanded into the *Puiseux series* (the series in fractional powers of ε).

Exercises

1. Let $A = \text{diag}(\lambda_1, \ldots, \lambda_n)$ where λ_i are real, pairwise distinct values. Assume that the matrix $A(\varepsilon) = A + A_1 \varepsilon$ is Hermitian and diag $(A_1) = 0$. Prove that
$$\lambda_i(\varepsilon) = \lambda_i + O\left(\varepsilon^2\right).$$

2. Assume that λ is a simple eigenvalue of a matrix A and that p and q are the corresponding left and right eigenvectors such that $q^T p = 1$. Prove that the perturbed matrix $A(\varepsilon) = A + A_1 \varepsilon$, for all sufficiently small ε, has a simple eigenvalue of the form
$$\lambda(\varepsilon) = \lambda + q^T A_1 p \varepsilon + O\left(\varepsilon^2\right).$$

3. If one of the condition numbers of the eigenvalues is large, then there exists at least one more that is large. Explain why.

4. Let A have simple eigenvalues with the condition numbers $s_1, \ldots, s_n$. Prove that if P is the eigenvector matrix, then $\text{cond}_2 P \geq \max_{1 \leq i \leq n} s_i$.

5. Let A have simple eigenvalues with the condition numbers $s_1, \ldots, s_n$. Prove that $s_1 = \ldots = s_n = 1$ if and only if the matrix A is normal.

6. Assume that A is diagonalizable by means of P:
$$P^{-1} A P = \text{diag}(\lambda_1, \ldots, \lambda_n),$$

and let $A + F$ be a perturbed matrix. Consider the discs

$$B_i = \left\{ z : \ |z - \lambda_i| \leq \|P^{-1}FP\|_2 \right\}, \quad i = 1, \ldots, n.$$

Let M be a union of m discs B_i, and assume that M does not intersect with the other discs. Prove that there are exactly m eigenvalues of $A + F$ located in M.

7. Assume that all the elements of A are different from zero. Then any eigenvalue $\lambda \in \lambda(A)$ is either an internal point of the region $R_1 \cup \ldots \cup R_n$ or the common boundary point for all the Gerschgorin discs $R_1, \ldots, R_n$.

8. In some sense, the unity matrix is "close" to a matrix of rank 1. Prove that, for any $\varepsilon > 0$, there exists a lower triangular matrix L such that $\|L\|_2 \leq \varepsilon$ and $I + L$ is a lower triangular part of some matrix of rank 1.

Lecture 5

5.1 Spectral distances

The statements of the corollaries of the Bauer–Fike theorem are "nonsymmetric" with respect to A and $A+F$: the matrix $A+F$, in contrast to A, can be nondiagonalizable or have different orders of Jordan blocks. We might be interested in "symmetric" theorems, which estimate some distance between the spectra of matrices.

The *Hausdorff distance* between A and B with the eigenvalues $\{\lambda_i\}$ and $\{\mu_j\}$, respectively, is defined as

$$\mathrm{hd}(A,\ B) \equiv \max\{\max_i \min_j |\lambda_i - \mu_j|, \max_j \min_i |\lambda_i - \mu_j|\}.$$

The *spectral p-distance* is defined as

$$d_p\,(A,\ B) \equiv \min_P \|\lambda(A) - P\lambda(B)\|_p,$$

where the minimum is taken over all the permutational matrices P and

$$\lambda(A) = [\lambda_1, \ldots, \lambda_n]^T, \quad \lambda(B) = [\mu_1, \ldots, \mu_n]^T.$$

5.2 "Symmetric" theorems

Theorem 5.2.1 (Elsner).

$$\mathrm{hd}\,(A,\ B) \le (\|A\|_2 + \|B\|_2)^{1-\frac{1}{n}} \|A - B\|_2^{\frac{1}{n}}.$$

Proof. Let $\mu \in \lambda(B)$ and $\lambda(A) = \{\lambda_i\}$. Suppose that the vectors $x_1, \ldots, x_n$ are the columns of the unitary matrix X, and $Bx_1 = \mu x_1$. Then

$$\prod_{i=1}^n |\lambda_i - \mu| = |\det\left((A - \mu I)\,X\right)|$$

(remember the Hadamar inequality: the determinant in modulus does not exceed the product of the 2-norms of its columns $\Leftrightarrow$ the volume of an n-dimensional cuboid does not exceed the product of the lengths of its edges)

$$\leq \prod_{i=1}^{n} \|(A - \mu I) x_i\|_2 \leq \|(A - B) x_1\|_2 \prod_{i=2}^{n} \|(A - \mu I) x_i\|_2$$

$$\leq \|A - B\|_2 (\|A\|_2 + \|B\|_2)^{n-1}. \quad \Box$$

Theorem 5.2.2 (Ostrowski–Elsner).

$$d_\infty(A, B) \leq (2n - 1) (\|A\|_2 + \|B\|_2)^{1 - \frac{1}{n}} \|A - B\|_2^{\frac{1}{n}}.$$

Proof. Consider the discs

$$D_i = \{z : |z - \lambda_i| \leq \mathrm{hd}(A, B)\} \quad \text{and} \quad D_i(\tau) = \{z : |z - \lambda_i| \leq \varepsilon(\tau)\},$$

where

$$\varepsilon(\tau) \equiv \left(2 \max_{0 \leq t \leq 1} \|A + t(B - A)\|_2 \right)^{1 - \frac{1}{n}} \|\tau(B - A)\|^{\frac{1}{n}}.$$

According to Theorem 5.2.1, all eigenvalues of $A + \tau(B - A)$ belong to a union of the discs $D_i(\tau)$. By analogy with the Gerschgorin discs, if some m discs $D_i(\tau)$ are isolated from the other discs, then their union contains exactly m eigenvalues of the matrix $A + \tau(B - A)$. $\Rightarrow$ If m discs $D_i(1)$ are isolated from the other discs, then the union of the corresponding discs D_i keeps exactly m eigenvalues μ_i. (Why?) Suppose that

$$\mu_i \in \bigcup_{1 \leq k \leq m} D_k, \quad k = 1, \ldots, m.$$

Then (prove this!) $|\mu_i - \lambda_j| \leq (2m - 1) \mathrm{hd}(A, B)$, $1 \leq i, j \leq m$. $\quad \Box$

5.3   Hoffman–Wielandt theorem

Theorem 5.3.1 (Hoffman–Wielandt) *For any normal matrices A and B,*

$$d_2(A, B) \leq \|A - B\|_F.$$

Proof. Since A and B are normal, by means of unitary matrices P and Q we transform them into diagonal matrices

$$D_A = \mathrm{diag}(\lambda_i) = P^* A P, \qquad D_B = \mathrm{diag}(\mu_i) = Q^* B Q.$$

Set $Z \equiv P^* Q$. Allowing for the unitary invariance property of the Frobenius norm, we find that

$$
\begin{aligned}
\|A - B\|_F^2 &= \|D_A - Z D_B Z^*\|_F^2 \\
&= \mathrm{tr}(D_A - Z D_B Z^*)(D_A - Z D_B Z^*)^* \\
&= \|D_A\|_F^2 + \|D_B\|_F^2 - 2 \operatorname{Re} \mathrm{tr}(Z D_B Z^* D_A^*).
\end{aligned}
$$

Further (check it),

$$\gamma \equiv 2\,\mathrm{Re}\;\mathrm{tr}(ZD_BZ^*D_A^*) = \sum_{i=1}^{n}\sum_{j=1}^{n}s_{ij}\alpha_{ij},$$

$$s_{ij} = |z_{ij}|^2, \quad \alpha_{ij} = 2\,\mathrm{Re}\;\lambda_i^*\mu_j.$$

For any fixed α_{ij}, a functional $\gamma = \gamma(S)$ is linear on the space of matrices $S = [s_{ij}]$. The values we are after are included in the set of values on those matrices S with nonnegative elements for which any column and row sum of elements is equal to 1. Such matrices S are called *doubly stochastic*.

The Birkhoff theorem. *Any doubly stochastic matrix S can be written as a convex combination of the permutational matrices P_k, $k = 1,\ldots,m$ (how many and which matrices themselves depend on S):*

$$S = \sum_{k=1}^{m}\nu_k P_k, \quad \nu_1 + \ldots + \nu_m = 1, \quad \nu_k \geq 0 \quad \forall\, k.$$

It is easy to verify that the set of all doubly stochastic matrices is compact. By the Birkhoff theorem,

$$\gamma(S) \leq \max_{1 \leq k \leq m}\gamma(P_k),$$

i.e., the maximal value of the function γ is attained at some permutational matrix. Denote it by Π. Then

$$
\begin{aligned}
||A - B||_F^2 &= ||D_A||_F^2 + ||D_B||_F^2 - 2\,\mathrm{Re}\;\mathrm{tr}(ZD_BZ^*D_A^*) \\
&\geq ||D_A||_F^2 + ||\Pi D_B\Pi^*||_F^2 - 2\,\mathrm{Re}\;\mathrm{tr}(\Pi D_B\Pi^*D_A^*) \\
&= ||D_A - \Pi D_B\Pi^*||_F^2 \geq d_2(A, B). \quad \square
\end{aligned}
$$

The next section is for those who want to know how the Birkhoff theorem can be proved.

5.4 Permutational vector of a matrix

For a matrix A, a permutational vector corresponding to a permutational matrix P means the vector made up of the diagonal elements of $\mathrm{diag}\,(P^T A)$. The locations of the elements extracted from A coincide with those of units in P.

The Birkhoff theorem is almost evident for those who accept as evident that any doubly stochastic matrix contains a nonzero permutational vector. As a matter of fact, if ν_1 is the minimal component of a nonzero permutational vector corresponding the permutational matrix P_1, then $S - \nu_1 P_1 = \phi_1 S_1$, where $0 \leq \nu_1, \phi_1 \leq 1$, $\nu_1 + \phi_1 = 1$, and S_1 is a doubly stochastic matrix which houses at least one zero element more than the original matrix S. Proceed by induction.

Attempt to prove the existence of a nonzero permutational vector. This is not at all simple. Prior to being able to do this we need to be armed with the following (nontrivial) theorem.

The Hall theorem. *Every permutational vector of an $n \times n$ matrix A contains a zero component if and only if A has a $p \times q$ zero submatrix with $p + q > n$.*

Necessity. Assume that every permutational vector of A contains zero. Write

$$A = \begin{bmatrix} a_{11} & \cdots & a_{1\,n-1} & a_{1n} \\ & & & a_{2n} \\ & B & & \cdots \\ & & & a_{nn} \end{bmatrix}.$$

If the matrix A is not zero then, without loss of generality, let $a_{1n} \neq 0$. Then the matrix B of order $n - 1$ must have zero in every permutational vector. Assume that (the induction in n) for B, there exists a $p \times q$ nonzero submatrix with $p + q > n - 1$. $\Rightarrow$ There exists a zero submatrix with $p + q = n$. Suppose it lies in the lower-left corner:

$$A = \begin{bmatrix} A_1 & * \\ 0 & A_2 \end{bmatrix}, \quad A_1 \in \mathbb{C}^{q \times q}, \ A_2 \in \mathbb{C}^{p \times p}.$$

At least one of the matrices A_1 and A_2 should have a zero on every permutational vector. Let it be A_1. By the inductive hypothesis, A_1 contains an $r \times s$ zero submatrix with $r + s > q$. Without loss of generality, assume that this submatrix is located in the lower-left corner of the matrix A_1. Then in the same corner of the matrix A, we have a $(r+p) \times s$ zero submatrix with $(r+p)+s = (r+s)+p > q+p = n$, just what was required to be proved.

Sufficiency. Assume that $p \leq q$ and a nonzero submatrix occupies the lower-left corner of A. Then

$$A = \begin{bmatrix} A_1 & * \\ 0_{q \times q} & A_2 \end{bmatrix},$$

and the last row of A_1 is zero. Thus every permutational vector is bound to capture some element of this row (i.e., zero). $\square$

Corollary. *Any doubly stochastic matrix contains a nonzero permutational vector.*

Proof. If not, by the Hall theorem, the doubly stochastic matrix (say, A) has a $p \times q$ zero submatrix with $p + q = n$. Suppose it occupies the lower-left corner:

$$A = \begin{bmatrix} A_1 & R \\ 0_{q \times q} & A_2 \end{bmatrix}.$$

The sum of all elements of R is equal to 0 (why?) $\Rightarrow$ $R = 0 \Rightarrow$ the matrices A_1 and A_2 are both doubly stochastic. Proceed by induction. $\square$

5.5   "Unnormal" extension

In the late 80s, an interesting generalization of the Hoffman–Wielandt theorem appeared on arbitrary diagonalizable matrices (not necessarily normal).

Theorem 5.5.1 (Sun–Zhang) *Assume that A and B are diagonalizable and P and Q are the corresponding eigenvector matrices. Then*

$$d_2(A, B) \leq \text{cond}_2(P) \, \text{cond}_2(Q) \, ||A - B||_F,$$

where

$$\text{cond}_2(P) \equiv ||P^{-1}||_2 ||P||_2, \quad \text{cond}_2(Q) \equiv ||Q^{-1}||_2 ||Q||_2.$$

Proof.

$$
\begin{aligned}
||A - B||_F &= ||PD_A P^{-1} - QD_B Q^{-1}||_F \\
&= ||P \left[D_A(P^{-1}Q) - (P^{-1}Q)D_B \right] Q^{-1}||_F \\
&\geq \frac{1}{||P^{-1}||_2 ||Q||_2} ||D_A Z - Z D_B||_F, \quad \text{where } Z \equiv P^{-1}Q.
\end{aligned}
$$

Consider the singular value decomposition $Z = V\Sigma U^*$. Then

$$||D_A Z - Z D_B||_F = ||V\{(V^* D_A V)\Sigma - \Sigma(U^* D_B U)\}U^*||_F = ||M\Sigma - \Sigma N||_F,$$

where the matrices $M = V^* D_A V$ and $N = U^* D_B U$ are, obviously, normal. By the Hoffman–Wielandt theorem,

$$d_2(A, B) = d_2(D_A, D_B) \leq ||M - N||_F.$$

Therefore, it is sufficient to make certain that

$$||M\Sigma - \Sigma N||_F \geq \sigma_{\min} ||M - N||_F.$$

Set $\Omega = \Sigma - \sigma_{\min} I$. Then (keep in mind that M and N are normal)

$$
\begin{aligned}
||M\Sigma - \Sigma N||_F^2 &= ||(M\Omega - \Omega N) + \sigma_{\min}(M - N)||_F^2 \\
&\leq ||M\Omega - \Omega N||_F^2 + \sigma_{\min}^2 ||M - N||_F^2 \\
&\quad + \sigma_{\min} \text{tr} \, \Omega\{(M - N)(M - N)^* + (M - N)^*(M - N)\}. \quad \square
\end{aligned}
$$

For a more detailed treatment of the above and related topics, I would recommend the nice book by Stewart and Sun.[1]

---

[1] G. W. Stewart and J. Sun. *Matrix Perturbation Theory*. Academic Press, 1990.

5.6   Eigenvalues of Hermitian matrices

Theorem 5.6.1 (Courant–Fischer) *Let* $\lambda_1 \geq \ldots \geq \lambda_n$ *be the eigenvalues of a Hermitian matrix* A *of order* n. *Then*

$$\lambda_k = \max_{\dim L = k} \min_{\substack{x \in L \\ x \neq 0}} \frac{x^* A x}{x^* x} \qquad (5.6.1)$$

and

$$\lambda_k = \min_{\dim L = n-k+1} \max_{\substack{x \in L \\ x \neq 0}} \frac{x^* A x}{x^* x}. \qquad (5.6.2)$$

Proof. Let $u_1, \ldots, u_n$ be the orthonormal eigenvectors corresponding to $\lambda_1, \ldots, \lambda_n$, respectively. As is easily seen,

$$x = \sum_{i=1}^{k} \xi_i u_i \quad \Rightarrow \quad \frac{x^* A x}{x^* x} \geq \lambda_k.$$

It follows that the right-hand side in (5.6.1) cannot be less than λ_k. (Why?)

Now consider a k-dimensional space M. There exists a nonzero vector $z \in M \cap K$, $K \equiv \mathrm{span}\,\{u_{n-k+1}, \ldots, u_n\}$. For any such a vector,

$$\frac{z^* A z}{z^* z} \leq \max_{\substack{x \in K \\ x \neq 0}} \frac{x^* A x}{x^* x} \leq \lambda_k.$$

Therefore, the right-hand side in (5.6.1) cannot be greater than λ_k. The proof of (5.6.2) is similar.   □

5.7   Interlacing properties

Theorem 5.7.1 *Suppose that* A *is a Hermitian matrix of order* n *and* B *is a leading principal submatrix of order* $n-1$. *Then the eigenvalues* $\lambda_1 \geq \ldots \geq \lambda_n$ *of* A *and the eigenvalues* $\mu_1 \geq \ldots \geq \mu_{n-1}$ *of* B *satisfy the following interlacing property:*

$$\lambda_k \geq \mu_k \geq \lambda_{k+1}, \quad k = 1, \ldots, n-1. \qquad (5.7.1)$$

Proof. Denote by $L_0 \subset \mathbb{C}^n$ any subspace of the vectors $x \in \mathbb{C}^n$ of the form

$$x = \begin{bmatrix} \hat{x} \\ 0 \end{bmatrix}.$$

Then any m-dimensional space of the vectors $\hat{x}$ corresponds to some m-dimensional space L_0 of the vectors x. In line with (5.6.1),

$$\mu_k = \max_{\dim L_0 = k} \min_{\substack{x \in L_0 \\ x \neq 0}} \frac{x^* A x}{x^* x} \leq \max_{\dim L = k} \min_{\substack{x \in L \\ x \neq 0}} \frac{x^* A x}{x^* x} = \lambda_k.$$

In line with (5.6.2),

$$\mu_k = \min_{\dim L_0 = (n-1)-k+1} \max_{\substack{x \in L_0 \\ x \neq 0}} \frac{x^* A x}{x^* x} \geq \min_{\dim L = n-k} \max_{\substack{x \in L \\ x \neq 0}} \frac{x^* A x}{x^* x} = \lambda_{k+1}. \quad \square$$

Theorem 5.7.2 *Suppose that a Hermitian matrix $A \in \mathbb{C}^{n \times n}$ is perturbed so that*

$$B = A + \varepsilon \, pp^*, \qquad \varepsilon \in \mathbb{R}, \qquad p \in \mathbb{C}^n, \, \|p\|_2 = 1, \tag{5.7.2}$$

and denote by $\lambda_1 \geq \ldots \geq \lambda_n$ and $\mu_1 \geq \ldots \geq \mu_n$ the eigenvalues of A and B, respectively. If $\varepsilon \geq 0$, then $\mu_1 \geq \lambda_1 \geq \mu_2 \geq \ldots \geq \mu_n \geq \lambda_n$, while if $\varepsilon \leq 0$, then $\lambda_1 \geq \mu_1 \geq \lambda_2 \geq \ldots \geq \lambda_n \geq \mu_n$. In either case, for all k,

$$\mu_k = \lambda_k + t_k \varepsilon, \qquad t_k \geq 0, \qquad and \quad t_1 + \ldots + t_n = 1. \tag{5.7.3}$$

Proof. Let $\varepsilon \leq 0$. By the Courant–Fischer theorem,

$$\lambda_k = \max_{\dim L = k} \min_{\substack{x \in L \\ x \neq 0}} \frac{x^* A x}{x^* x} \leq \max_{\dim L = k} \min_{\substack{x \in L \\ x \neq 0}} \frac{x^* B x}{x^* x} = \mu_k,$$

and, further, for $k \geq 2$,

$$\mu_k \leq \max_{\dim L = k} \min_{\substack{x \in L, \, x \perp p \\ x \neq 0}} \frac{x^* B x}{x^* x} = \max_{\dim L = k} \min_{\substack{x \in L, \, x \perp p \\ x \neq 0}} \frac{x^* A x}{x^* x}$$

$$\leq \max_{\dim L = k-1} \min_{\substack{x \in L \\ x \neq 0}} \frac{x^* A x}{x^* x} = \lambda_{k-1}.$$

Thus, we may write $\mu_k = \lambda_k + t_k \varepsilon$ with $t_k \geq 0$. Adding these equations, we obtain $\operatorname{tr} B = \operatorname{tr} A + \varepsilon (t_1 + \ldots + t_n)$ while the starting equation (5.7.2) implies $\operatorname{tr} B = \operatorname{tr} A + \varepsilon$. Hence, $t_1 + \ldots + t_n = 1$. The proof for $\varepsilon \leq 0$ is similar. $\quad \square$

5.8   What are clusters?

Sometimes the most of (though not all) eigen or singular values are massed near some set on the complex plane (one or several points, as a rule). Such a set is said to be a cluster. However, we need to bring far more into the picture.

Consider a sequence of matrices $A_n \in \mathbb{C}^{n \times n}$ with eigenvalues $\lambda_i(A_n)$ and a subset M of complex numbers. For any $\varepsilon > 0$, denote by $\gamma_n(\varepsilon)$ the number of those eigenvalues A_n that fall outside the ε distance from M. Then M is called a *(general) cluster* if

$$\lim_{n \to \infty} \frac{\gamma_n(\varepsilon)}{n} = 0 \quad \forall \varepsilon > 0,$$

and a *proper cluster* if

$$\gamma_n(\varepsilon) \leq c(\varepsilon) \quad \forall n, \quad \forall \varepsilon > 0.$$

We consider chiefly the clusters consisting of one or several (a finite number of) points ($M = \mathbb{C}$ is never of interest).

One might also be interested in the singular value clusters. To distinguish between the eigen and the singular value cases, let us write $\gamma_n(\varepsilon; \lambda)$ and $\gamma_n(\varepsilon; \sigma)$, respectively. For brevity, we denote that M is a cluster by $\lambda(A_n) \sim M$ or $\sigma(A_n) \sim M$.

Of course, we tacitly assume that A_n are the elements of some common process (for instance, they might arise from a digitization of some operator equation on a sequence on grids).

5.9 Singular value clusters

To prove that a sequence has a cluster, we can try to find a "close" sequence for which this is already established. That "closeness" can be treated in a rather broad sense.[2]

Theorem 5.9.1 *Suppose that A_n and B_n are such that*

$$\|A_n - B_n\|_F^2 = o(n) \tag{5.9.1}$$

or, alternatively,

$$\operatorname{rank}(A_n - B_n) = o(n). \tag{5.9.2}$$

In either case, any singular value cluster for A_n is also a singular value cluster for B_n, and vice versa.

Proof. First, consider the case (5.9.1). Denote by $\sigma_1(A_n) \geq \ldots \geq \sigma_n(A_n)$ and $\sigma_1(B_n) \geq \ldots \geq \sigma_n(B_n)$ the singular values of A and B. Then it follows from the Hoffman–Wielandt theorem (prove this) that

$$\sum_{k=1}^{n}(\sigma_k(A_n) - \sigma_k(B_n))^2 \leq \|A_n - B_n\|_F^2.$$

Take an arbitrary $\delta > 0$ and denote by $\alpha_n(\delta)$ the number of those $k \in \{1, \ldots, n\}$ for which $|\sigma_k(A_n) - \sigma_n(B_n)| \geq \delta$. Together with (5.9.1) the above implies that

$$\alpha_n(\delta)\,\delta^2 = o(n) \qquad \Rightarrow \qquad \alpha_n(\delta) = o(n).$$

Consider a subset $M \subset \mathbb{C}$, and denote by $\gamma_n^A(\varepsilon)$ and $\gamma_n^B(\varepsilon)$ the $\gamma_n(\varepsilon; \sigma)$ functions (introduced in the previous section) for A_n and B_n, respectively. Choosing $\delta = \varepsilon/2$ we obtain (why?)

$$\gamma_n^B(\varepsilon) \leq \gamma_n^A\left(\frac{\varepsilon}{2}\right) + \alpha_n\left(\frac{\varepsilon}{2}\right).$$

Since the right-hand side is $o(n)$, that will do the proof.

[2]E. E. Tyrtyshnikov. A unifying approach to some old and new theorems on distribution and clustering. *Linear Algebra Appl.* 323: 1–43 (1996).

In the case of (5.9.2), we have to recollect that the singular values in question are the square roots of the eigenvalues of the Hermitian matrices $A_n^* A_n$ and $B_n^* B_n$. Obviously, $\operatorname{rank}(A_n^* A_n - B_n^* B_n) = o(n)$. Thus, do not loose generality if we assume that A_n and B_n are both Hermitian. If so, we may apply the interlacing property stated in Theorem 5.7.2. Doing this several times (as many as $\operatorname{rank}(A_n - B_n)$ is) we arrive in the end at

$$\gamma_n^B(\varepsilon) \leq \gamma_n^A(\varepsilon) + \operatorname{rank}(A_n - B_n). \quad \Box$$

5.10 Eigenvalue clusters

To state that the eigenvalue clusters for A_n and B_n coincide, we ought to add something to the premises of Theorem 5.9.1. For example, we can formulate the following.

Theorem 5.10.1 *Suppose that* $A_n, B_n \in \mathbb{C}^{n \times n}$ *are diagonalizable for every* n, *and denote by* P_n, Q_n *the corresponding eigenvector matrices. If*

$$\operatorname{cond}_2^2 P_n \operatorname{cond}_2^2 Q_n \, \|A_n - B_n\|_F^2 \; = \; o(n),$$

then any eigenvalue cluster for A_n *is also an eigenvalue cluster for* B_n, *and vice versa.*

Proof. Using the "unnormal" extension of the Hoffman–Wielandt theorem discussed above, we follow the same lines as in proving Theorem 5.9.1. $\Box$

Some stronger statements can be made if one of the sequences is a constant matrix, for example, the zero one.[3] (In print.)

Theorem 5.10.2 *Suppose* A_n *has a singular value cluster at zero and, in addition, is bounded so that, uniformly in all sufficiently small* $\varepsilon > 0$,

$$\log \|A_n\|_2 = O\left(\frac{n}{\gamma_n(\varepsilon; \sigma)}\right).$$

Then A_n *has the eigenvalue cluster at zero, too.*

Proof. Let $|\lambda_1(A_n)| \geq \ldots \geq |\lambda_n(A_n)|$ and $\sigma_1(A_n) \geq \ldots \geq \sigma_n(A_n)$ be the eigenvalues in modulus and the singular values of A_n, respectively. We use the following Weyl inequality:

$$\prod_{k=1}^m |\lambda_k(A_n)| \; \leq \; \prod_{k=1}^m \sigma_k(A_n), \qquad m = 1, \ldots, n. \tag{5.10.1}$$

(Prove this. Hint: it is not too hard a task for us, because we already wield the Hoffman–Wielandt theorem and interlacing properties.)

---

[3]E. E. Tyrtyshnikov and N. L. Zamarashkin. On eigen and singular value clusters, *Calcolo* (1997).

By contradiction, suppose the eigenvalues of A_n are not clustered at 0. Therefore, there exist $\varepsilon_0, c_0 > 0$ and some subset of increasing indices $\mathcal{N} = \{n_1, n_2, \dots\}$ such that $\gamma_n(\varepsilon_0; \lambda) \geq c_0 n \quad \forall n \in \mathcal{N}$. Without loss of generality, assume that $n_k = k \quad \forall k$. Choose any $\varepsilon > 0$. Using (5.10.1), we obtain

$$\varepsilon_0^{\gamma_n(\varepsilon_0;\lambda)} \leq \prod_{k=1}^{\gamma_n(\varepsilon_0;\lambda)} |\lambda_k(A_n)| \leq \prod_{k=1}^{\gamma_n(\varepsilon_0;\lambda)} \sigma_k(A_n) \leq \|A_n\|_2^{\gamma_n(\varepsilon;\sigma)} \varepsilon^{\gamma_n(\varepsilon_0;\lambda)-\gamma_n(\varepsilon;\sigma)}$$

$$\Rightarrow \quad \left(\frac{\varepsilon_0}{\varepsilon}\right)^{\frac{\gamma_n(\varepsilon_0;\lambda)}{n}} \leq \left(\frac{\|A_n\|_2}{\varepsilon}\right)^{\frac{\gamma_n(\varepsilon;\sigma)}{n}}.$$

By the contradictory assumption, if $\varepsilon < \varepsilon_0$, then the left-hand side is lower-bounded by a positive constant. For sufficiently small ε, this constant can be made arbitrary large, and in particular, larger than the upper bound of the right-hand side. $\square$

Exercises

1. Suppose that $A, B \in \mathbb{C}^{n \times n}$ are Hermitian with eigenvalues $\lambda_1 \geq \dots \geq \lambda_n$ and $\mu_1 \geq \dots \geq \mu_n$. Prove that $\sum_{i=1}^{n} |\lambda_i - \mu_i|^2 \leq \|A - B\|_F^2$.

2. Suppose that $A, B \in \mathbb{C}^{n \times n}$ have the singular values $\lambda_1 \geq \dots \geq \lambda_n$ and $\mu_1 \geq \dots \geq \mu_n$. Prove that $\sum_{i=1}^{n} |\lambda_i - \mu_i|^2 \leq \|A - B\|_F^2$.

3. Suppose that $A \in \mathbb{C}^{n \times n}$ is Hermitian and $B = P^* A P$, where $P \in \mathbb{C}^{k \times n}$ has orthonormal columns. Prove that

$$\lambda_{n-k+1}(A) + \dots + \lambda_k(A) \leq \operatorname{tr} B \leq \lambda_1(A) + \dots + \lambda_k(A).$$

4. Prove that $\min_{\operatorname{rank} B \leq k} \|A - B\|_F = \sqrt{\sigma_{n-k+1}^2(A) + \dots + \sigma_n^2(A)}$.

5. Consider an upper bidiagonal $n \times n$ matrix A_n with 1 on the main diagonal and 2 on the neighboring one. Prove that

$$\sigma_1(A_n), \dots, \sigma_{n-1}(A_n) \in [1, 3] \quad \text{and} \quad \sigma_n(A_n) \in (0, 2^{1-n}).$$

6. For any $A \in \mathbb{C}^{n \times n}$, prove the Weyl inequality

$$\prod_{k=1}^{m} |\lambda_k(A_n)| \leq \prod_{k=1}^{m} \sigma_k(A_n), \quad m = 1, \dots, n.$$

7. Produce a sequence of matrices A_n for which the eigenvalues are clustered at some point, but the singular values are not clustered at any point.

8. Produce a sequence of matrices A_n for which the singular values are clustered at 1, but the eigenvalues are not.

Lecture 6

6.1 Floating-Point numbers

There are only finite many numbers in a computer. These are the so-called *floating-point numbers*:

$$a = \pm \left(\frac{d_1}{p} + \frac{d_2}{p^2} + \ldots + \frac{d_t}{p^t} \right) \cdot p^\alpha.$$

Here, $p, \alpha, d_1, \ldots, d_t$ are integers. The number $p > 0$ is said to be the base of a computer arithmetic. The number in brackets is the mantissa, and α is the exponent of a floating-point a. The numbers $d_i \in \{0, 1, \ldots, p-1\}$ are termed digits, and t is the length of the mantissa. As a rule, $d_1 \neq 0$. After all, there are some integers, L and U, which are the bounds of α: $L \leq \alpha \leq U$. A special floating-point number is $a = 0$.

Thus, the set of all floating-point numbers is determined by the parameters p, t, L, and U.

6.2 Computer arithmetic axioms

When fed into a computer and operated on as floating-point numbers, the quantities go through a rounding-off as usual. A rounding-off is a mapping of the real numbers into the floating-point numbers. Let $fl(x)$ denote the rounding-off result for x. Then the following axiom applies:

$$fl(x) = x(1 + \varepsilon), \tag{6.2.1}$$

where $|\varepsilon| \leq \eta$ as long as $fl(x) \neq 0$. We *define* η as the lowest upper bound for $|\varepsilon|$. Then for the school rule of rounding off,[1] we obtain (check it)

$$\eta = \frac{1}{2} p^{1-t}. \tag{6.2.2}$$

---

[1] It reads: given a number to be rounded off, take a number nearest to it with a prescribed mantissa; in the case of two candidates, take the largest one.

By tradition, the result of a computer operation $*$ on floating-point numbers a and b is signified by $fl(a * b)$. We postulate that if $fl(a * b) \neq 0$, then

$$fl(a * b) = a * b(1 + \varepsilon), \qquad |\varepsilon| \leq \eta. \tag{6.2.3}$$

This is the primary axiom we rely on when studying the roundoff errors in numerical algorithms.

NB: The relative error ε must be small only when the result of a computer operation is not zero.

Sometimes numbers are rounded off by cutting off the "superfluous" digits. In this case, the equality $x = a * b$ does not necessarily entail that $fl(a * b) = fl(x)$. For example, suppose that $p = 2$, $t = 2$. Let $a = 0.11$, $b = 0.0001$ and $x = a - b = 0.1011$. Then $fl(x) = 0.10$ while $fl(a - b) = 0.11$ (the handling of numbers reduces to that of t-digit ones on a special device called a "summator").

For the cutting-off rule, $\eta = p^{1-t}$.

6.3   Roundoff errors for the scalar product

Exact equations for actually computed quantities include a good many different $\varepsilon_1, \varepsilon_2, \ldots$. Not to overload the equations, let us designate all these $\varepsilon_1, \varepsilon_2, \ldots$ by the same letter ε. Adopt the notation

$$(1 + \varepsilon)^n \equiv \prod_{i=1}^{n} (1 + \varepsilon_{k_i});$$

if such things occur several times, then all ε_{k_i} in the corresponding expansions are regarded as different. This causes no problem when deriving some inequalities, because any ε is subject to (6.2.3) (so long as no floating-point zero pops up).

Now, let $\tilde{\alpha}$ be an actually computed scalar product:

$$\alpha = x^T y, \qquad x = [x_1, \ldots, x_n]^T, \qquad y = [y_1, \ldots, y_n]^T \in \mathbb{R}^n.$$

Suppose the scalar product appears after the following prescriptions:

$$\alpha = 0; \quad \text{DO } i = 1, n$$
$$\alpha = \alpha + x_i y_i$$
$$\text{END DO}$$

Then, in chime with (6.2.3), we find that

$$\tilde{\alpha} = \sum_{i=1}^{n} x_i y_i (1 + \varepsilon)^{n+1-i}. \tag{6.3.1}$$

6.4   Forward and backward analysis

One can interpret (6.3.1) in different ways. In the spirit of the *forward analysis*, we are to estimate the deviation between the exact and computed answers:

$$|\tilde{\alpha} - \alpha| \le n\eta |x^T| \|y| + \mathcal{O}(\eta^2). \tag{6.4.1}$$

We assume above and from now on that if $A = [a_{ij}]$ then $|A| = [|a_{ij}|]$.

In the spirit of the *backward analysis*, we are to represent a really computed answer as the result of exact computations with perturbed data and, then, to derive a bound on the corresponding (termed *equivalent*) perturbation:

$$\tilde{\alpha} = \tilde{x}^T \tilde{y},$$
$$|\tilde{x} - x| \le \tfrac{1}{2}n\eta|x| + \mathcal{O}(\eta^2), \quad |\tilde{y} - y| \le \tfrac{1}{2}n\eta|y| + \mathcal{O}(\eta^2). \tag{6.4.2}$$

Clearly, the perturbations can be distributed between x and y in some other ways.

6.5   Some philosophy

In the course of examining the roundoffs, it is typical to ignore the quantities of order $\mathcal{O}(\eta^2)$. Following Wilkinson (one of the prominent experts of the field), we emphasize that the main objective is not the derivation of neat bounds ("the bound itself is usually the least important part of it") but mostly to expose (and to fix, if possible) the potential instabilities of an algorithm at issue. It would be not right to think that a large inaccuracy in an answer is due to a large number of operations (and hence, roundoffs). More frequently, there is just one operation that spoils the picture.

6.6   An example of "bad" operation

An operation with a bad reputation is that of subtracting close numbers of the same sign. This operation, as such, is not any worse than others, for its own roundoff error is of order η. Yet it can *amplify dramatically* the previously accumulated errors. In particular, let $\tilde{a} \approx a$, $\tilde{b} \approx b$; then

$$fl(\tilde{a} - \tilde{b}) = (\tilde{a} - \tilde{b})(1 + \varepsilon) = (a - b)(1 + \varepsilon + \delta),$$

where

$$\delta = \left(\frac{(\tilde{a} - a) - (\tilde{b} - b)}{a - b} \right)(1 + \varepsilon).$$

Obviously, δ can get very large.

6.7   One more example

Some routine was used to compute the eigenvectors of a block triangular matrix with 2×2 blocks, and, after its run, it was seen that exactly one eigenvector had a residual of three orders beyond those of the other eigenvectors. In this case, the instability originated from the solution of a homogeneous system with two equations and two unknowns:

$$a_1 x_1 + a_2 x_2 = 0,$$
$$b_1 x_1 + b_2 x_2 = 0;$$

$$|x_1| = 1 \text{ or } |x_2| = 1.$$

Assume that the vectors $a = [a_1, \ a_2]^T$ and $b = [b_1, \ b_2]^T$ are nonzero and approximately collinear; set $|a_1 b_2 - a_2 b_1| = \delta$. If we take

$$x_1 = -a_2/\|a\|_\infty, \qquad x_2 = a_1/\|a\|_\infty, \tag{$*$}$$

then the residuals are

$$r_1 \equiv |a_1 \tilde{x}_1 + a_2 \tilde{x}_2| \le 2\eta \|a\|_\infty,$$
$$r_2 \equiv |b_1 \tilde{x}_1 + b_2 \tilde{x}_2| \le \frac{\delta}{\|a\|_\infty} + 2\eta \|b\|_\infty. \tag{6.7.1}$$

Alternatively, if we take

$$x_1 = -b_2/\|b\|_\infty, \qquad x_2 = b_1/\|b\|_\infty, \tag{$**$}$$

then

$$r_1 \le \frac{\delta}{\|b\|_\infty} + 2\eta \|a\|_\infty,$$
$$r_2 \le 2\eta \|b\|_\infty. \tag{6.7.2}$$

The bounds (6.7.1) and (6.7.2) differ in that one contains $\delta/\|a\|_\infty$ while the other contains $\delta/\|b\|_\infty$. If $\|b\|_\infty$ exceeds $\|a\|_\infty$ by three orders, then the residual for the first approach may be, roughly speaking, three orders greater. Thus, we can recommend the following recipe: if $\|a\|_\infty \ge \|b\|_\infty$, then choose $(*)$; otherwise, $(**)$.

6.8   Ideal and machine tests

As a rule, an algorithm undergoes some testing, when the computed answers are compared with the known exact ones. Shortsighted programmers might see no discrepancy between the ideal and machine tests, which leads them, sometimes, to wrong conclusions.

For example, a routine that solves linear systems by the Gauss elimination method was tried on a test $Ax = b$ with the *Hilbert matrix*

$$A = \left[\frac{1}{i+j-1} \right]_{ij=1}^n$$

and the solution vector $x = [1, \ldots, 1]$. The right-hand side was computed by multiplying A by x. For $n = 10$, the computed solution $\tilde{x}$ (in the "REAL*8" regime on FORTRAN) turned out to be

$$||\tilde{x} - x||_\infty \approx 0.9 \cdot 10^2 .$$

For all this, do not hurry to say that the algorithm produced a "bad" solution.

In actual fact, instead of the exact A and b the computer had gotten some close but different $\hat{A}$ and $\hat{b}$, and thus the machine test was different from the ideal one. The Hilbert matrix is ill-conditioned; consequently, the exact solution $\hat{x}$ for the machine test may go (and did) far from x:

$$\hat{x} \approx \begin{bmatrix} 1.042595644 \\ 0.459944616 \\ 1.284169655 \\ 2.772926997 \\ 2.217756963 \\ 3.252956378 \\ -5.410267887 \\ -46.120499977 \\ 93.504857996 \\ -43.063313904 \end{bmatrix} .$$

It is clear, now, that the computed solution $\tilde{x}$ should be compared with $\hat{x}$ but no longer with x. Should this be done, the relative errors in the components of the solution do not exceed 10^{-8}.

6.9 Up or down

The most spectacular example of an unstable algorithm I ever encountered is the following.[2] We are to compute the integrals

$$E_n = \int_0^1 x^n e^{x-1} \, dx, \quad n = 1, \ldots .$$

A simple recursive formula emerges immediately after integrating by parts. Using it, we obtain the following "down-to-up" algorithm:

$$E_1 = 1/e; \quad E_n = 1 - n \, E_{n-1}, \quad n = 2, 3, \ldots . \tag{*}$$

It is easy to prove that (do it)

$$0 < E_n < \frac{1}{n+1}.$$

[2]G. E. Forsythe, M. A. Malcolm, and C. B. Moler. *Computer Methods for Mathematical Computations*. Prentice-Hall, Englewood Cliffs, NJ, 1977.

Yet, when running this algorithm, you shall see very large and even negative values of E_n. The problem is that even a small inaccuracy in E_1 is multiplied by $n!$ in E_n.

Curiously, if we start with various initial data

$$E_1 = c; \quad E_n = 1 - n E_{n-1}, \quad n = 2, 3, \ldots,$$

then (prove this)

$$\lim_{n \to \infty} E_n = \left\{ \begin{array}{ll} 0 & , \quad c = 1/\mathrm{e}, \\ \infty & , \quad \text{otherwise.} \end{array} \right.$$

A good idea is to begin with a sufficiently large N and run the same but reversed recursive formula. Thus, we obtain the following "up-to-down" algorithm:

$$E_N = 1; \quad E_n = (1 - E_{n+1})/n, \quad n = N - 1, \ N - 2, \ldots, 1. \qquad (**)$$

Now, any inaccuracy in E_N will be diminished just next to a few initial steps. For example, the algorithm $(**)$ can be used to compute $\mathrm{e} = 2.718281828\ldots$ to machine precision.

6.10   Solving the triangular systems

For solving a system $Lx = b$ with a lower triangular matrix $L = [\, l_{ij} \,]$, we can apply the following *forward substitution method*:

$$\mathrm{DO}\ i = 1,\ n$$
$$x_i = \left(b_i - \sum_{j=1}^{i-1} l_{ij} x_j \right) / l_{ii}$$
$$\mathrm{END\ DO}$$

If $\tilde{x}_i$ is a really computed value then we obtain

$$\tilde{x}_i = \left(b_i - \sum_{j=1}^{i-1} l_{ij} \tilde{x}_j (1 + \varepsilon)^{i-j} \right) (1 + \varepsilon)^2 / l_{ii}.$$

Set

$$\tilde{l}_{ij} = \left\{ \begin{array}{ll} l_{ij}(1 + \varepsilon)^{i-j}, & i > j, \\ l_{ii}/(1 + \varepsilon)^2, & i = j, \\ 0, & i < j. \end{array} \right.$$

Then

$$\tilde{x}_i = \left(b_i - \sum_{j=1}^{i-1} \tilde{l}_{ij} \tilde{x}_j \right) / \tilde{l}_{ii},$$

i.e., $\tilde{x}_i$ is a component of the exact solution of a system with the same right-hand side but a perturbed coefficient matrix $\tilde{L} = \left[\tilde{l}_{ij} \right]$.

How close $\tilde{L}$ and L are is easy to estimate:

$$\left|\tilde{l}_{ij} - l_{ij}\right| \leq \begin{cases} |l_{ij}| (i-j)\eta + \mathcal{O}(\eta^2), & i > j, \\ |l_{ii}| 2\eta + \mathcal{O}(\eta^2), & i = j, \\ 0, & i < j. \end{cases} \qquad (6.10.1)$$

Thus, the following holds:

Theorem 6.10.1 *For the forward substitution method, a really computed solution $\tilde{x}$ of a system $Lx = b$ satisfies a perturbed system $\tilde{L}\tilde{x} = b$, where $\tilde{L}$ is a lower triangular matrix such that*

$$\left|\tilde{L} - L\right| \leq n\eta|L| + \mathcal{O}(\eta^2). \qquad (6.10.2)$$

The inequality (6.10.2) is a concise but rougher form of (6.10.1). The result we have gotten is practically ideal for an algorithm from the viewpoint of the backward analysis of roundoffs.

The case of the *backward substitution method* is treated similarly.

Exercises

1. Is it true that $fl(\frac{a+b}{2}) \in [a, b]$?

2. To find e^x for $x = -13$, somebody sums up the series

$$e^x = 1 + x + \frac{x^2}{2!} + \frac{x^3}{3!} + \dots$$

and gets an enormous error. Why? Would you suggest a remedy?

3. The back substitution method is know to provide always a very small residual. Explain why. Write a routine for this method, and run it to solve the system

$$\begin{bmatrix} 1 & 2 & & & \\ & 1 & 2 & & \\ & & \ddots & \ddots & \\ & & & 1 & 2 \\ & & & & 1 \end{bmatrix}_{n \times n} \begin{bmatrix} x_1 \\ \vdots \\ \\ x_n \end{bmatrix} = \begin{bmatrix} 1 \\ \vdots \\ 1 \\ 1/3 \end{bmatrix}$$

for $n = 40$. How small is the residual? To what accuracy is the solution?

4. Devise an algorithm for summing up n real numbers providing the following properties:

$$fl(a_1 + \dots + a_n) = \tilde{a}_1 + \dots + \tilde{a}_n,$$

where

$$|\tilde{a}_i - a_i| \leq \eta \log_2 n |a_i| + \mathcal{O}(\eta^2).$$

Lecture 7

7.1 Direct methods for linear systems

We build up algorithms using some set of elementary operations. If a problem is solved with a finite number of elementary operations (in exact arithmetic), then the corresponding method is said to be *direct*.

Not every problem can be given a direct method. For instance, it is impossible to solve the equation $x^2 = 2$ using a finite number of arithmetic operations.

For all this, a direct method exists as soon as the square root is relegated to an elementary operation. All the same, we are aware (due to Galois and Abel) that, even with the roots of any degree, there can be no direct method for finding zeroes of algebraic polynomials of degree five and larger. That is why one should not try to invent a direct method to compute eigenvalues. (Why?)

For solving linear systems, direct methods exist using only arithmetic operations; sometimes the square root is added to them.

We shall discuss, in some detail, the classical direct methods for *dense unstructured* matrices. These are the Gauss method and other elimination methods making zeroes by the Givens (rotation) or the Householder (reflection) transformations. These methods are related to the LU and QR decompositions for the coefficient matrix.

7.2 Theory of the LU decomposition

A square matrix A is called *strongly regular* if all its leading submatrices (A itself as well) are nonsingular.

By the LU decomposition of A, the equality $A = LU$ is understood, where L is a lower unitriangular matrix (i.e., triangular with units along the main diagonal) while U is a nonsingular upper triangular matrix.

Theorem 7.2.1 *A matrix A admits the LU decomposition if and only if it is strongly regular.*

Proof. The "only if" part is evident. For the "if" part we use induction. Write

$$A = \begin{bmatrix} a & c^T \\ b & D \end{bmatrix}.$$

Then

$$\begin{bmatrix} 1 & 0 \\ -\frac{1}{a}b & I \end{bmatrix} \begin{bmatrix} a & c^T \\ b & D \end{bmatrix} = \begin{bmatrix} a & c^T \\ 0 & A_1 \end{bmatrix}, \qquad A_1 \equiv D - \frac{1}{a}bc^T.$$

It is easy to see that the matrix A_1 is also strongly regular. By the inductive assumption, it admits the LU decomposition $A_1 = L_1 U_1$. Set

$$L = \begin{bmatrix} 1 & 0 \\ \frac{1}{a}b & L_1 \end{bmatrix}, \qquad U = \begin{bmatrix} a & c^T \\ 0 & U_1 \end{bmatrix}$$

and calculate

$$LU = \begin{bmatrix} a & c^T \\ b & L_1 U_1 + \frac{1}{a}bc^T \end{bmatrix} = \begin{bmatrix} a & c^T \\ b & D \end{bmatrix} = A. \quad \square$$

Corollary 7.2.1 *The LU decomposition is determined uniquely.*

The equality $L_1 U_1 = L_2 U_2$ implies $L_2^{-1} L_1 = U_2 U_1^{-1} \equiv D$. The products of lower triangular matrices and the inverse of such a matrix are lower triangular matrices. The same holds for upper triangular matrices. Consequently, D is simultaneously lower and upper triangular $\Rightarrow$ it is diagonal. Since $L_2^{-1} L_1$ is a unitriangular matrix, we obtain $D = I$. $\square$

Corollary 7.2.2 *All leading minors of A are positive if and only if U has positive diagonal elements.*

Corollary 7.2.3 *If a strongly regular matrix $A \in \mathbb{C}^{n \times n}$ is symmetric ($A = A^T$), then it admits the LDL^T decomposition $A = LDL^T$, where L is lower unitriangular while D is a nonsingular diagonal matrix.*

Set $D = \operatorname{diag}(U)$. Then $A = LDD^{-1}U = A^T = D^{-1}U^T(DL^T)$. Obviously, the matrix $D^{-1}U^T$ is lower unitriangular. By the uniqueness of the LU decomposition, $DL^T = U$. $\square$

Corollary 7.2.4 *If a strongly regular matrix $A \in \mathbb{C}^{n \times n}$ is Hermitian ($A = A^*$), then it admits the LDL^* decomposition $A = LDL^*$, where L is lower unitriangular while D is a nonsingular diagonal matrix.*

Prove this.

Let C denote a lower triangular matrix with positive diagonal elements. The decomposition $A = CC^*$ is called *the Cholesky decomposition*. The decomposition $A = CC^T$ is said to be the symmetric triangular decomposition.

Theorem 7.2.2 *For a matrix $A \in \mathbb{C}^{n \times n}$ to have the symmetric triangular decomposition (the Cholesky decomposition), it is necessary and sufficient that it be symmetric (Hermitian) with positive leading minors.*

Proof. Necessity is evident. To prove sufficiency, we take up the decomposition $A = LDL^T$ (or $A = LDL^*$) and set $C = L\text{diag}\left(\sqrt{d_1}, \ldots, \sqrt{d_n}\right)$, where $D = \text{diag}(d_1, \ldots, d_n)$. $\square$

7.3 Roundoff errors for the *LU* decomposition

The proof of the theorem on the *LU* decomposition was constructive. It contains a (recursive) algorithm that everybody knows by the name Gauss algorithm.

Theorem 7.3.1 *Assume A is a machine, real, strongly regular matrix of order n. Then the actually computed by the Gauss algorithm matrices $\tilde{L}$ and $\tilde{U}$ satisfy the following inequality:*

$$\left|\tilde{L}\tilde{U} - A\right| \le 3(n-1)\eta\left(|A| + |\tilde{L}||\tilde{U}|\right) + \mathcal{O}(\eta^2), \qquad (7.3.1)$$

provided that the machine zero has not occurred in any elementary operation.

The theorem can be proved by induction.[1] This is one of the most remarkable results of the roundoff analysis for algebraic algorithms.

7.4 Growth of matrix entries and pivoting

The bound 7.3.1 would be smart, if only without the term $|\tilde{L}||\tilde{U}|$. One might speculate that it is probably not a tight bound. Yet, the term $|\tilde{L}||\tilde{U}|$ points, by any chance, to the critical bottleneck of the Gauss algorithm, a potential growth of entries of the triangular factors.

The growth of matrix entries could be very dangerous. Let p and t denote the base of machine arithmetic and the length of the mantissa. Let $n = 2$, and consider

$$A \equiv \begin{bmatrix} a & c \\ b & d \end{bmatrix} = \begin{bmatrix} p^{-t} & 1. \\ 1. & 1. \end{bmatrix}.$$

Then

$$\tilde{L} = \begin{bmatrix} 1. & 0 \\ p^t & 1. \end{bmatrix}, \qquad \tilde{U} = \begin{bmatrix} p^{-t} & 1. \\ 0 & -p^t \end{bmatrix}$$

and, therefore,

$$\tilde{L}\tilde{U} - A = \begin{bmatrix} 0 & 0 \\ 0 & 1. \end{bmatrix}.$$

---

[1]G. H. Golub and C. F. Van Loan. *Matrix Computations*. The Johns Hopkins University Press, 1989.

An awfully large inaccuracy is accounted for by a small value of a, *the pivot*.

Apparently, we should look for a better pivot, or, in other words, we need *pivoting*. For example, any entry of the column at work can be made a pivot by swapping rows. A sane choice is the entry maximal in modulus. This column pivoting ensures that all the entries of L are upper bounded ($|l_{ij}| \le 1$). At the same time,

$$\rho \equiv \frac{\max_{i,j} |u_{ij}|}{\max_{i,j} |a_{ij}|} \le 2^{n-1}. \tag{7.4.1}$$

If we pick up as a pivot the maximal in modulus entry with the minimal row index, then the above estimate for the growth coefficient ρ is attained on the following matrix

$$A = \begin{bmatrix} 1 & & & & 1 \\ -1 & 1 & & 0 & 1 \\ -1 & -1 & \ddots & & 1 \\ \vdots & \vdots & & \ddots & \vdots \\ -1 & -1 & \cdots & -1 & 1 \end{bmatrix}. \tag{7.4.2}$$

It might be a comfort to note that one must be lucky enough to observe a growth this large in actual practice.

7.5 Complete pivoting

Once k steps of the Gauss method are done, we can try as a pivot (by swapping rows and columns) any nonzero entry of the "active submatrix" A_k

Let the pivots be maximal in modulus among the entries of each active submatrix. For this *complete pivoting*, there was a long-standing hypothesis due to Wilkinson that $\rho \le n$ for real matrices. In 1991 this hypothesis was found to be untrue (N. Gould produced an example of real matrix of order 13 with $\rho > 13$).[2]

---

[2]N.Gould. On growth in Gaussian elimination with complete pivoting. *SIAM J. Matrix Anal. Appl.* 12 (2): 354–361 (1991).

The most radical means for preventing the growth of entries is to make zeroes by orthogonal transformations.

7.6 The Cholesky method

Let A be a real symmetric matrix with positive leading minors. Take $n = 3$ and try to satisfy the equation

$$
\begin{bmatrix} a_{11} & a_{21} & a_{31} \\ a_{21} & a_{22} & a_{32} \\ a_{31} & a_{32} & a_{33} \end{bmatrix} = \begin{bmatrix} c_{11} & & 0 \\ c_{21} & c_{22} & \\ c_{31} & c_{32} & c_{33} \end{bmatrix} \begin{bmatrix} c_{11} & c_{21} & c_{31} \\ & c_{22} & c_{32} \\ 0 & & c_{33} \end{bmatrix}. \Rightarrow
$$

$$
\begin{array}{lll}
c_{11} = \sqrt{a_{11}}, & c_{21} = a_{21}/c_{11}, & c_{31} = a_{31}/c_{11}; \\
c_{22} = \sqrt{a_{22} - c_{21}^2}, & c_{32} = (a_{32} - c_{31}c_{21})/c_{22}; & \\
c_{33} = \sqrt{a_{33} - c_{31}^2 - c_{32}^2}. & &
\end{array}
$$

Here is how the Cholesky algorithm looks for an arbitrary n:

DO $k = 1$, n

$$
c_{kk} = \left(a_{kk} - \sum_{j=1}^{k-1} c_{kj}^2 \right)^{1/2} \tag{$*$}
$$

DO $i = k + 1$, n

$$
c_{ik} = \left(a_{ik} - \sum_{j=1}^{k-1} c_{ij} c_{kj} \right) / c_{kk} \tag{$**$}
$$

END DO
END DO

Suppose the expressions $(*)$ and $(**)$ are computed so that the relative error ε for any arithmetic operation and the square root operation have the estimate $|\varepsilon| \leq \eta$. Then for the actually computed quantities $\tilde{c}_{ij}$, we find that

$$
\tilde{c}_{kk}^2 / (1 + \varepsilon)^3 = a_{kk} - \sum_{j=1}^{k-1} \tilde{c}_{kj}^2 (1 + \varepsilon)^{k-j} \qquad \text{for } (*);
$$

$$
\tilde{c}_{ik} \tilde{c}_{kk} / (1 + \varepsilon)^2 = a_{ik} - \sum_{j=1}^{k-1} \tilde{c}_{ij} \tilde{c}_{kj} (1 + \varepsilon)^{k-j} \quad \text{for } (**).
$$

From the first equality,

$$
\sqrt{\sum_{j=1}^{k} |\tilde{c}_{kj}|^2} \leq \sqrt{a_{kk}} + \mathcal{O}(\eta) \leq \sqrt{a_{kk}} + \mathcal{O}(\eta).
$$

Therefore,

$$
\left| \sum_{j=1}^{k} \tilde{c}_{ij} \tilde{c}_{kj} - a_{ik} \right| \;\leq\; \eta\,(k+1) \sum_{j=1}^{k} |\tilde{c}_{ij}|\,|\tilde{c}_{kj}| \;+\; \mathcal{O}(\eta^2)
$$

$$
\leq\; \eta\,(k+1)\sqrt{\sum_{j=1}^{k} |\tilde{c}_{ij}|^2}\sqrt{\sum_{j=1}^{k} |\tilde{c}_{kj}|^2} \;+\; \mathcal{O}(\eta^2)
$$

$$
\leq\; \eta\,(n+1)\sqrt{\sum_{j=1}^{k} |\tilde{c}_{ij}|^2}\sqrt{\sum_{j=1}^{i} |\tilde{c}_{kj}|^2} \;+\; \mathcal{O}(\eta^2)
$$

$$
\leq\; \eta\,(n+1)\,\sqrt{a_{ii}}\,\sqrt{a_{kk}} \;+\; \mathcal{O}(\eta^2).
$$

Finally,

$$
|\tilde{C}\,\tilde{C}^T - A| \leq \eta\,(n+1)
\begin{bmatrix} \sqrt{a_{11}} \\ \cdots \\ \sqrt{a_{nn}} \end{bmatrix}
[\sqrt{a_{11}}, \ldots, \sqrt{a_{nn}}] \;+\; \mathcal{O}(\eta^2).
$$

For the Cholesky method, thus, we are free from any trouble with the growth of entries.

7.7 Triangular decompositions and linear systems solution

The whole of the solution process for $Ax = b$ with a nonsingular A can consist of the following three stages:

1) $A = LU$ (computing the LU decomposition).

2) $Ly = b$ (forward substitution).

3) $Ux = y$ (backward substitution).

The first stage is the hardest, for the Gauss method requires $\frac{2}{3}n^3 + \mathcal{O}(n^2)$ arithmetic operations (check this). In the common case, we advocate column pivoting. The second and the third stages require $n^2 + \mathcal{O}(n)$ operations each.

On a parallel computer, the time distribution among the three stages can differ. Let us imagine an abstract parallel computer with an arbitrarily large number of processors and instantaneous exchange of data between all devices. Then, for each of the three stages, there are algorithms that work in $\mathcal{O}(\log_2^2 n)$ parallel steps. Nobody knows whether it is possible to cut back this time even if only for triangular systems.

7.8 How to refine the solution

Suppose we have an approximate solution $\tilde{x}_0$ computed by the Gauss method. The following recipe shows how to refine it:

1) Compute the residual $r_{i-1} = A\tilde{x}_{i-1} - b$.

2) Solve a system for the correction vector c_{i-1}: $Ac_{i-1} = r_{i-1}$.

3) $x_i = \tilde{x}_{i-1} - c_{i-1}$.

If the above prescriptions are performed in exact arithmetic, then the vector x_i must be the exact solution for $Ax = b$. In the case of roundoffs, we obtain a new approximation $\tilde{x}_i$ to the exact solution x. Will $\tilde{x}_i$ be any better than $\tilde{x}_{i-1}$?

Assume that the actually computed and the exact correction vectors ($\tilde{c}_i$ and c_i) are bonded by the inequality

$$\|\tilde{c}_i - c_i\|_2 \le \vartheta \|c_i\|_2, \tag{7.8.1}$$

where $0 < \vartheta < 1/(1 + \eta)$. Then, for the vectors $c_i = A^{-1}(A\tilde{x}_i - b) = \tilde{x}_i - x$,

$$\|c_i\| \le q\|c_{i-1}\|_2 + \eta\|x\|_2, \qquad q = \vartheta(1 + \eta). \tag{7.8.2}$$

It follows easily that

$$\|c_i\|_2 \le q^i \|c_0\|_2 + \left(q^{i-1} + q^{i-2} + \ldots, +1\right)\eta\|x\|_2.$$

Hence,

$$\frac{\|c_i\|_2}{\|x\|_2} \le \frac{\eta}{1 - q} + O(q^i). \tag{7.8.3}$$

Thus, if the approximation we start from is not too bad (i.e., (7.8.1) is fulfilled) then the above refinement method allows us to approach the solution with an accuracy of about $\eta/(1 - q)$.

Since the initial approximation can have a fairly small residual, it is usually recommended to

(a) compute the residual r_{i-1} in a higher precision regime and only then round it off to the working precision, and

(b) normalize the residual by a factor of p^k, where p is the base of machine arithmetic, prior to the correction vector computation, i.e., to solve the system $A\hat{c}_{i-1} = r_{i-1}/\|r_{i-1}\|$, and, finally, assign $c_{i-1} = \hat{c}_{i-1}\|r_{i-1}\|$.

We regard the refinement process as cheap, because, at this stage, when solving the system for the correction vector, we know that the LU decomposition of A is already computed.

Generally speaking, the refinement process allows us to achieve good accuracy even for ill-conditioned matrices. It makes sense if the machine coefficient matrix and right-hand side are regarded as *exact*. Otherwise, the refinement process is not a panacea to fight against the ill condition. At any rate, its application to such problems needs specific substantiation.

Exercises

1. Let $\mathcal{M}$ denote a set of $n \times n$ matrices of the form $A = \alpha I - N$, where N has nonnegative entries and all its eigenvalues in modulus are less than α. Prove that any $A \in \mathcal{M}$ admits the LU decomposition, and what is more, $L \in \mathcal{M}$ and $U \in \mathcal{M}$.

2. Show that the Gauss method with column pivoting is equivalent (in exact and machine arithmetic) to the Gauss method without pivoting applied to the same matrix but with swapped rows.

3. Assume that a system $Ax = b$ was solved through the three stages according to Section 7.7, and there were no machine zeroes in the run. Let the actually computed quantities be $\tilde{L}$, $\tilde{U}$ and $\tilde{x}$. Prove that $(A + E)\tilde{x} = b$, where

$$|E| \le n\eta \left(4|A| + 5|\tilde{L}||\tilde{U}| \right) + \mathcal{O}(\eta^2).$$

4. Suppose that the LU decomposition is computed by the Gauss method without pivoting for a matrix $A \in \mathbb{R}^{n \times n}$ which is row-wise diagonally dominant. Prove that in this case the growth factor

$$\rho \equiv \frac{\max\limits_{i,j} |u_{ij}|}{\max\limits_{i,j} |a_{ij}|}$$

   does not exceed 2.

5. Suppose that the LU decomposition is computed by the Gauss method without pivoting for a symmetric positive definite matrix $A \in \mathbb{R}^{n \times n}$. Prove that the growth factor is equal to 1.

6. If A is a Hermitian positive definite matrix, then the spectral condition number of the active submatrix on each step of the Gauss method does not exceed the spectral condition number of A. Prove this.

7. Assume that, in the Cholesky method, the expressions (*) and (**) are computed so that the relative error ε has the estimate $|\varepsilon| \le \eta$. Prove that[3]

$$\|\tilde{C}\tilde{C}^T - A\|_F \lesssim 2\eta \|A\|_F.$$

8. Devise a parallel algorithm that solves a nonsingular triangular system through $\mathcal{O}(\log_2^2 n)$ parallel steps. (Hint: this problem is not too difficult!)

9. The Cholesky algorithm as we put in Section 7.6 requires $O(n^2)$ parallel steps. Find a version of it that runs through $O(n)$ parallel steps.

---

[3]V. V. Voevodin. *Computational Bases of Linear Algebra.* Nauka, Moscow, 1977 (in Russian).

Lecture 8

8.1 The QR decomposition of a square matrix

Given a square matrix A, a decomposition $A = QR$, where Q is unitary and R is upper triangular, is said to be the QR decomposition of A. In contrast to the LU decomposition, one does not fear a growth of entries in the case of the QR decomposition. (Why?)

Theorem 8.1.1 *The QR decomposition exists for any square matrix.*

Proof. Assume A is nonsingular $\Rightarrow$ A^*A is positive definite $\Rightarrow$ all leading submatrices of A^*A are positive definite $\Rightarrow$ all leading minors in A^*A are positive $\Rightarrow$ there exists the Cholesky decomposition $A^*A = R^*R$ (R is upper triangular) $\Rightarrow$ the matrix $Q \equiv AR^{-1}$ is unitary:

$$Q^*Q = (AR^{-1})^*(AR^{-1}) = R^{-*}(A^*A)R^{-1} = (R^{-*}R^*)(RR^{-1}) = I.$$

If A is singular, then, for all sufficiently large n, the perturbed matrix $A_n = A + \frac{1}{n}I$ must be nonsingular (why?). Hence, it admits the QR decomposition $A_n = Q_n R_n$. The set of unitary matrices is compact (why?) $\Rightarrow$ there exists a convergent subsequence

$$Q_{n_k} \to Q \quad \Rightarrow \quad Q_{n_k}^* A \to Q^*A \equiv R.$$

As is easily seen, the matrix Q is unitary whereas R is upper triangular. $\square$

Corollary. *For a nonsingular A, the matrices Q and R are determined uniquely provided that the main diagonal of R is positive.*

8.2 The QR decomposition of a rectangular matrix

Let $A \in \mathbb{C}^{m \times n}$ and $m \geq n$. Then there is a decomposition $A = QR$, where R is a square upper triangular matrix of order n whereas Q has orthonormal columns.

One can prove this by embedding A into a square matrix with zeroes in the place of uncertain entries.

8.3   Householder matrices

A matrix $H = H(u) = I - 2uu^*$, where $||u||_2 = 1$, is called a *Householder matrix*, or a reflection matrix. Verify that

(a) H is unitary;

(b) H is Hermitian;

(c) $Hu = -u$, and $Hv = v \quad \forall \; v \perp u$.

Lemma 8.3.1 *For arbitrary vectors* $a, b \in \mathbb{C}^n$ *of the same length, there exist a scalar* γ *and a Householder matrix* H *such that*

$$Ha = \gamma b, \quad |\gamma| = 1.$$

Proof. If $H = H(u)$, then

$$a - 2(u^*a)u = \gamma b.$$

If a and b are nonzero collinear vectors, then we can set $u = a/||a||_2$. Otherwise, we take

$$u = \frac{a - \gamma b}{||a - \gamma b||_2}$$

and choose γ so that

$$2(u^*a) = ||a - \gamma b||_2^2 \quad \Leftrightarrow$$

$$2(a^*a - \gamma^*b^*a) = ||a - \gamma b||_2^2 = ||a||_2^2 + ||b||_2^2 - 2\operatorname{Re}(\gamma^*b^*a).$$

Since $||a||_2 = ||b||_2$, we obtain $\gamma^*b^*a = \operatorname{Re}(\gamma^*b^*a) \; \Leftrightarrow$ the quantity $\gamma^* b^*a$ is real. If $b^*a = 0$ then we can take any γ such that $|\gamma| = 1$. Otherwise, there are two options:

$$\gamma = b^*a/|b^*a|, \quad \text{or} \quad \gamma = -b^*a/|b^*a|. \quad \square$$

8.4   Elimination of elements by reflections

According to Lemma 8.3.1, for any column $a \in \mathbb{C}^n$, there is a Householder matrix H such that

$$Ha = \gamma \, [||a||_2, 0, \ldots, 0]^T, \quad |\gamma| = 1.$$

In this case, H is determined by the vector $u = v/||v||_2$, where

$$v = [a_1 - \gamma \, ||a||_2, a_2, \ldots, a_n]^T.$$

If $a_1 \neq 0$, then we recommend choosing

$$\gamma = -a_1/|a_1|.$$

(There will be no subtraction of equal sign numbers!)

For any matrix $A \in \mathbb{C}^{n \times n}$, there exist Householder matrices $H_1, \ldots, H_{n-1}$ such that

$$H_{n-1} \ldots H_1 A = R$$

is an upper triangular matrix.

Let H_i be determined by a vector u_i with zeroes in its first $i-1$ components; H_i must annihilate subdiagonal entries of the ith column of the matrix $H_{i-1} \ldots H_1 A$.

Since the product of unitary matrices $Z \equiv H_{n-1} \ldots H_1$ is still unitary, $A = Q R$, $Q = Z^*$, — one more (constructive) proof of the existence of the QR decomposition.

For the QR decomposition to be done via reflections, we perform $\frac{4}{3} n^3 + \mathcal{O}(n^2)$ arithmetic operations (prove this). For comparison: in the Gauss method for the LU decomposition, we need $\frac{2}{3} n^3 + \mathcal{O}(n^2)$ arithmetic operation.

8.5 Givens matrices

A matrix $G_{kl} \in \mathbb{R}^{n \times n}$ is called a *rotation matrix*, or Givens matrix, if it differs from the unity matrix by only a 2×2 submatrix of the form

$$M(\phi) = \begin{bmatrix} \cos \phi & -\sin \phi \\ \sin \phi & \cos \phi \end{bmatrix},$$

located in the rows and columns k and l.

Check that G_{kl} is an orthogonal matrix.

8.6 Elimination of elements by rotations

If a vector $[a_1, a_2]^T \in \mathbb{R}^2$ is nonzero, then the choice

$$\cos \phi = \frac{-a_1}{\sqrt{a_1^2 + a_2^2}}, \quad \sin \phi = \frac{a_2}{\sqrt{a_1^2 + a_2^2}}$$

is to eliminate its second component:

$$M(\phi) \begin{bmatrix} a_1 \\ a_2 \end{bmatrix} = \begin{bmatrix} \alpha \\ 0 \end{bmatrix}.$$

When premultiplying A by a Givens matrix G_{kl}, we can nullify any one entry of rows k or l. Consequently, we can move from A to an upper triangular form R through a sequence of premultiplications by Givens matrices:

$$G_{n-1\,n} \ldots G_{1n} \ldots G_{13} G_{12} A = R.$$

For the QR decomposition to be done via rotations, we perform $2 n^3 + \mathcal{O}(n^2)$ arithmetic operations (prove this).

8.7   Computer realizations of reflections and rotations

A computer implementation of reflections or rotations provides for $\tilde{Q}$ and $\tilde{R}$ such that

$$\|A - \tilde{Q}\,\tilde{R}\| \leq c_1(n)\,\eta\,\|A\| + \mathcal{O}(\eta^2), \qquad (8.7.1)$$

$$\|\tilde{Q}^*\,\tilde{Q} - I\| \leq c_2(n)\,\eta + \mathcal{O}(\eta^2), \qquad (8.7.2)$$

where $c_1(n)$ and $c_2(n)$ are functions of n (depending upon the norms in question and the implementation details).

The reflection and rotation algorithms are far different from the standpoint of parallel computations. Using the so-called fan-in (pairwise) addition, we can find the QR decomposition by reflections in $\mathcal{O}(n \log n)$ parallel steps. Alternatively, the standard rotations with no touch-up are richer in (somewhat "hidden", from the first glance) parallel actions, for only $\mathcal{O}(n)$ parallel steps are required. (Check it!)

8.8   Orthogonalization method

One can find the QR decomposition with no rotations or reflections. Suppose that $n = 3$ and we wish to fulfill the equation $A = Q\,R$:

$$[a_1 \; a_2 \; a_3] = [q_1 \; q_2 \; q_3] \begin{bmatrix} r_{11} & r_{12} & r_{13} \\ 0 & r_{22} & r_{23} \\ 0 & 0 & r_{33} \end{bmatrix}.$$

For convenience, we introduce nonnormalized vectors p_1, p_2, and p_3 collinear to q_1, q_2, and q_3, respectively. The first columns coincide if we set $p_1 = a_1$, $r_{11} = \|p_1\|_2$, and $q_1 = p_1/r_{11}$. Then the equation for the second columns is

$$a_2 = q_1\,r_{12} + q_2\,r_{22}.$$

Since q_2 ought to be orthogonal to q_1, we premultiply both sides by q_1^* and find

$$r_{12} = q_1^* a_2, \qquad p_2 = a_2 - q_1\,r_{12},$$
$$r_{22} = \|p_2\|_2, \qquad q_2 = p_2/r_{22}.$$

Next, equate the third columns:

$$a_3 = q_1\,r_{13} + q_2\,r_{23} + q_3\,r_{33}.$$

Premultiply both sides by q_1^*, q_2^* and find

$$r_{13} = q_1^* a_3, \; r_{23} = q_2^* a_3,$$
$$p_3 = a_3 - q_1\,r_{13} - q_2\,r_{23}, \qquad r_{33} = \|p_3\|_2, \; q_3 = p_3/r_{33}.$$

What we have above is the classical *Gram-Schmidt orthogonalization algorithm*. It serves to obtain an orthonormal basis in a linear hull span $\{a_1, \ldots, a_k\}$

spanned by linearly independent vectors $a_1, \ldots, a_k$. Here is how the Gram-Schmidt looks in the general case:

$$p_j = a_j - \sum_{i=1}^{j-1} q_i \, (q_i^* a_j), \quad q_j = p_j / \|p_j\|_2, \quad j = 1, \ldots, k. \tag{8.8.1}$$

Here, $r_{ij} = q_i^* a_j$.

To orthogonalize n vector of size n, we perform $2\,n^3 + \mathcal{O}\,(n^2)$ arithmetic operations (check this).

8.9   Loss of orthogonality

In machine arithmetic, the linear span for the vectors $\tilde{q}_1, \ldots, \tilde{q}_k$ actually computed by formulas (8.8.1) coincides with the linear span for some perturbed vectors $a_1 + f_1, \ldots, a_k + f_k$. The perturbations $f_1, \ldots, f_k$ are guaranteed to be small. (Prove it!)

However, the vectors $\tilde{q}_1, \ldots, \tilde{q}_k$ can be far away from orthogonal ones. Let us try to realize why. It seems natural to measure the orthogonality property for the vectors $\tilde{q}_1, \ldots, \tilde{q}_i$ by a quantity

$$\delta_i \equiv \|\tilde{Q}_i^* \tilde{Q}_i - I\|_2, \qquad \tilde{Q}_i == [\tilde{q}_1, \ldots, \tilde{q}_i], \quad i = 1, \ldots, k. \tag{8.9.1}$$

Assume that the computations of the step $i + 1$ are free from roundoffs. Then

$$\begin{aligned}
\beta_{i+1} \equiv \tilde{Q}_i^* \tilde{q}_{i+1} &= \frac{1}{\|\tilde{p}_{i+1}\|_2} \tilde{Q}_i^* \left(a_{i+1} - \tilde{Q}_i \tilde{Q}_i^* a_{i+1} \right) \\
&= \frac{1}{\|\tilde{p}_{i+1}\|_2} \left(I - \tilde{Q}_i^* \tilde{Q}_i \right) \tilde{Q}_i^* a_{i+1}.
\end{aligned}$$

Hence,

$$\|\beta_{i+1}\|_2 \leq \delta_i \sqrt{1 + \delta_i} \, \frac{\|a_{i+1}\|_2}{\|\tilde{p}_{i+1}\|_2}. \tag{8.9.2}$$

An evident upper bound holds

$$\delta_{i+1} = \left\| \begin{bmatrix} \tilde{Q}_i^* \tilde{Q} - I & \beta_{i+1} \\ \beta_{i+1}^* & 0 \end{bmatrix} \right\|_2 \leq \delta_i + 2 \|\beta_{i+1}\|_2 \tag{8.9.3}$$

(quite sufficient for our purposes). A little neater bound is as follows:

$$\delta_{i+1} \leq \frac{\delta_i + \sqrt{\delta_i^2 + 4 \|\beta_{i+1}\|_2^2}}{2}. \tag{8.9.4}$$

To sum up,

$$\delta_{i+1} \leq \text{const} \, \frac{\|a_{i+1}\|_2}{\|\tilde{p}_{i+1}\|_2} \, \delta_i \sqrt{1 + \delta_i}.$$

Even if δ_i is small, we can not keep a rein on δ_{i+1} when $\|\tilde{p}_{i+1}\|_2$ becomes small!

8.10   How to prevent the loss of orthogonality

Assume that we have computed $\tilde{q}_1, \ldots, \tilde{q}_i$ which are still "sufficiently well" orthogonal (that is, δ_i is small). To maintain orthogonality in the step $i + 1$, we consider the following *reorthogonalization process*:

$$\begin{aligned} p^{(0)} &= a_{i+1}; \\ p^{(j)} &= (I - \tilde{Q}_i\,\tilde{Q}_i^*)\,p^{(j-1)}, \quad j = 1, 2, \ldots. \end{aligned}$$

The iteration for $j = 1$ pertains to the standard Gram-Schmidt algorithm. If we quit on the jth iteration, then, finally, $q_{i+1} = p^{(j)}/\|p^{(j)}\|_2$.

It is easy to see that $\beta^{(j)} \equiv \tilde{Q}_i^* p^{(j)} = (I - \tilde{Q}_i^* \tilde{Q}_i)^j\, \tilde{Q}_i^* p^{(0)}$.   So long as $\delta_i < 1$, the matrix $I - \tilde{Q}_i^* \tilde{Q}_i$ will be convergent $\Rightarrow \beta^{(j)} \to 0$ for $j \to \infty$.

W.Hoffmann advocates continuing the reorthogonalization unless

$$\frac{\|p^{(j)}\|_2}{\|p^{(j-1)}\|_2} \le \frac{1}{2}. \tag{$*$}$$

(In his paper,[1] one finds very useful speculations and numerical examples to support $(*)$. Still, I am not sure that there is a rigorously proved bound on the orthogonality measure for the strategy all through.)

8.11   Modified Gram-Schmidt algorithm

Consider the following two pseudoroutines based on formulas (8.8.1):

DO $\quad j = 1, k$	DO $\quad j = 1, k$
$\quad p_j = a_j$	$\quad p_j = a_j$
$\quad$ DO $i = 1, j - 1$	$\quad$ DO $i = 1, j - 1$
$\qquad p_j = p_j - q_i\,(q_i^*\,a_j)$	$\qquad p_j = p_j - q_i\,(q_i^*\,p_j)$
$\quad$ END DO	$\quad$ END DO
$\quad q_j = p_j/\|p_j\|_2$	$\quad q_j = p_j/\|p_j\|_2$
END $\quad$ DO	END $\quad$ DO

The second one is said to be the *modified Gram-Schmidt algorithm*. It comes from the former by a change of a_j onto p_j (in the innermost loop).

Both algorithms behave the same in exact arithmetic, but not on a machine. As A. Bjorck showed, the modified algorithm has the estimate

$$\delta_k \le c(n, k)\, \frac{\sigma_{\max}}{\sigma_{\min}}\, \eta,$$

where $\sigma_{\max}$ and $\sigma_{\min}$ are the maximal and minimal singular values of the $n \times k$ matrix $A = [a_1, \ldots, a_k]$, $n \ge k$. Thus, in contrast to the standard

---

[1]W.Hoffmann. Iterative algorithms for Gram–Schmidt orthogonalization. *Computing* 41: 335–348 (1989).

Gram–Schmidt algorithm, the modified algorithm is guaranteed to keep some orthogonality.

Recently, it was revealed that the modified Gram–Schmidt algorithm coincides literally (and so *in machine arithmetic*) with the Housholder algorithm for the QR decomposition of the following zeroes-on-top rectangular matrix

$$\hat{A} = \begin{bmatrix} 0_{k \times k} \\ A \end{bmatrix} .$$

In 1992, A. Bjorck and C. C. Paige[2] used this observation (due to C. Sheffield) to rederive the above Bjorck estimate.

If you are still not satisfied with the orthogonality provided by the modified algorithm, do not hesitate to touch it up by the reorthogonalization process.

8.12 Bidiagonalization

A matrix $B = [b_{ij}]$ is called (upper) *bidiagonal* if $b_{ij} = 0$ whenever $i > j$ or $i + 1 < j$.

Any $n \times n$ matrix A can be reduced to a bidiagonal form

$$B = P A Q$$

so that P and Q are finite products of the Householder (or, alternatively, Givens) matrices.

Suppose we use reflections. First, we multiply A from the left by a reflection matrix that cancels all the subdiagonal entries of the first column. Then, we multiply the result from the right by a reflection matrix that annihilates the first row's entries from 3 to n. It is important that keep safeguard all previously obtained zeroes in the first column!

Further, by premultiplication, we nullify the subdiagonal entries of the second column, and then, by postmultiplication, make zeroes in the second row's entries from 4 to n, and so on. In every step, we do not lose any previously obtained zeroes (check it).

By the unitary bidiagonalization, many (not all!) problems for a matrix A reduce to the same problems but for the bidiagonal matrix B. These are the most important examples:

- *The singular value decomposition of an arbitrary matrix A* can be easily constructed from the singular value decomposition of the corresponding bidiagonal matrix B.

- *The least squares problem* for any A, i. e., the minimization problem for $\|A x - b\|_2$ in x, can be tackled by changing variables $x = Q u$, $b = P f$, and, then, solving the minimization problem for $\|B u - f\|_2$ in u with the bidiagonal matrix B.

---

[2] A. Bjorck and C. C. Paige. Loss and recapture of orthogonality in the modified Gram–Schmidt algorithm. *SIAM J. Matrix Anal. Appl.* 13 (1): 176–190 (1992).

8.13   Unitary similarity reduction to the Hessenberg form

A matrix $H = [h_{ij}]$ is called (upper) *Hessenberg* if $h_{ij} = 0$ for $i > j + 1$.
Any $n \times n$ matrix A can be reduced to a unitary similar Hessenberg matrix

$$H = P^* A P$$

so that P is a product of a finite number of reflection (or rotations).
Suppose we use reflections. Consider a reflection matrix P_1 so that

$$P_1 [a_{11}, a_{21}, a_{31}, \ldots, a_{n1}]^T = [a_{11}, *, 0, \ldots, 0]^T .$$

When a matrix is premultiplied by P_1, nothing is done to its first row of entries.
Hence, when a matrix is postmultiplied by $P_1^* = P_1$, nothing happens to its
first column of entries. In $P_1 A P_1$, thus, we have zeroes in the first column
of entries from 3 to n. Step by step, we choose the reflections $P_2, \ldots, P_{n-2}$
so that the premultiplication by P_i makes zeroes in the ith column of entries
from $i + 2$ to n. Finally, $P = P_1 \ldots P_{n-2}$.

Any matrices that are similar have the same spectra. Consequently, if
the eigenvalues and eigenvectors are required for A, we know how to proceed
as soon as we have learned how to solve the same problem for a Hessenberg
matrix H. Important property: since P is unitary, the condition numbers for
simple eigenvalues for H remain the same as those for A (prove it).

If A is a Hermitian matrix, then the above Hessenberg matrix H is, in
fact, tridiagonal (prove it).

Exercises

1. Is it true that, for any vectors $a, b \in \mathbb{C}^n$ of equal length, there is a
   reflection matrix H such that $H a = b$?

2. Is it true that any rotation matrix can be expressed as a product of a
   finite number of reflection matrices ?

3. Is it true that any reflection matrix can be expressed as a product of a
   finite number of rotation matrices?

4. Let $H = H^*$. Prove the inequality

$$\left\| \begin{bmatrix} H & \beta \\ \beta^* & 0 \end{bmatrix} \right\|_2 \leq \frac{\|H\|_2 + \sqrt{\|H\|_2^2 + 4\|\beta\|_2^2}}{2} .$$

   (This is behind the estimate (8.9.4)).

5. Show that the modified Gram-Schmidt algorithm for the orthogonalization of columns of an $n \times k$ matrix A is equivalent, including roundoffs,

to the Householder algorithm for the QR decomposition applied to an augmented rectangular matrix of the form

$$\hat{A} = \left[\begin{array}{c} 0_{k \times k} \\ A \end{array} \right].$$

6. Let $H = P^* A P$, where P is a unitary matrix. Prove that, for any simple eigenvalue, its condition number for A is equal to that for H.

Lecture 9

9.1 The eigenvalue problem

"The eigenvalue problem" is a short name for a variety of settings and problems related to computing eigenvalues, eigenvectors, spectral projectors, etc.

Suppose we are to find all the eigenvalues of a matrix $A \in \mathbb{C}^{n \times n}$. How would we do this? A rather old idea is to find the characteristic polynomial's coefficients and, thus, reduce the problem to computing the polynomial roots.

There are several "efficient" algorithms for computing the characteristic polynomial's coefficients (a direct method exists that requires $\mathcal{O}(n^3)$ arithmetic operations). However, the idea was dismissed with the advent of computers, because the eigenvalues might be slightly sensitive to small perturbations of matrix entries but terribly sensitive to small perturbations of the characteristic polynomial's coefficients (in this way we risk reducing a "good" problem to a "bad" one).

9.2 The power method

Another old idea appeared more fruitful. Suppose that A has a basis of the eigenvectors: $A z_i = \lambda_i z_i$, and assume that

$$|\lambda_1| > |\lambda_2| > \ldots > |\lambda_n|.$$

Consider the following *power method*:

y_0 is an initial guess, $x_0 = y_0 / \|y_0\|$;
$y_k = A x_{k-1}$, $x_k = y_k / \|y_k\|$, $k = 1, 2, \ldots$.

(Why is the normalization necessary?)

If $y_0 = \mu_1 z_1 + \ldots + \mu_n z_n$ and $\mu_1 \neq 0$, then x_k is collinear to

$$\frac{A^k y_0}{\mu_1 \lambda_1^k} = z_1 + \mathcal{O}\left(\left|\frac{\lambda_2}{\lambda_1}\right|^k\right) \quad \Longrightarrow \quad x_k^* A x_k \longrightarrow \lambda_1.$$

Thus, the power method allows one to approximate the senior (greatest in modulus) eigenvalue.

To find λ_i for $i \neq 1$, one can try some α for $\lambda_i - \alpha$ to become the senior eigenvalue for $A - \alpha I$. Is this always possible?

There might be an alternative: to carry out one step of the inductive proof of the Schur theorem and come to a matrix of order $n-1$ with the eigenvalues $\lambda_2, \ldots, \lambda_n$ (to put it into action, one definitely proceeds rather implicitly; the method as such is referred to as the *deflation method*).

In its pure form, the power method is not used very often. However, most modern algorithms exploit the idea behind it.

9.3   Subspace iterations

The power method develops naturally into the *method of subspace iterations*:

Y_0 is an $n \times m$ matrix of rank m,

Y_0 $= X_0 R_0$ (the QR decomposition of a rectangular matrix);

Y_k $= A X_{k-1}$, and $Y_k = X_k R_k$ (the QR decomposition), $k = 1, 2, \ldots$.

The method generates a sequence of subspaces $L_k = \operatorname{im} X_k$. In this role, the QR decomposition is similar to the normalization in the power method (besides, it simply produces orthonormal bases in L_k).

If L_k is an A-invariant subspace (recall that it means that $A L_k \subset L_k$), then $A_k = X_k^* A X_k$ is a diagonal block of a block triangular matrix that is similar to A (explain this). That is why we infer that

$$\lambda(A_k) \subset \lambda(A).$$

Under some hypotheses, as k increases, the subspace L_k approaches some A-invariant subspace. Consequently, the eigenvalues of the $m \times m$ matrix A_k (usually m is far less than n) approximate some eigenvalues of A.

To get on to a precise analysis, we need to define the distance between subspaces.

9.4   Distances between subspaces

The distance between a vector $x \in \mathbb{C}^n$ and a subspace $M \subset \mathbb{C}^n$ is defined as

$$\rho(x, M) \equiv \min_{y \in M} \|x - y\|_2.$$

It suggests that the distance between subspaces L and M would have been defined as

$$\rho(L, M) \equiv \max_{x \in L, \|x\|_2 = 1} \rho(x, M).$$

Will ρ fulfil the axioms of a metric space?

We shall see shortly that it will if we consider only subspaces of equal dimension. If not, it might be that $\rho(L, M) \neq \rho(M, L)$. Verify that if L is not orthogonal to M and dim L > dim M, then

$$\rho(L, M) = 1, \quad \text{whereas} \quad \rho(M, L) < 1.$$

All the axioms of a metric space are obviously met if we define the distance as follows:

$$\text{dist}\,(L, M) \equiv \max\,\{\rho(L, M), \rho(M, L)\}.$$

9.5 Subspaces and orthoprojectors

A matrix P is termed an *orthoprojector* if $P^2 = P$ and $P^* = P$.

Proposition 1. *If $L = \operatorname{im} P$, then $P x$ is the orthogonal projection of the vector x onto the subspace L.*

Proof.
$$y \in L \Rightarrow y = P v \Rightarrow$$
$$y^*(x - P x) = v^* P^*(x - P x) = v^*(P x - P^2 x) = 0. \quad \Box$$

It is clear that there is a one-to-one correspondence between subspaces L and orthoprojectors P_L providing for $\operatorname{im} P_L = L$. (Prove this.)

Proposition 2. *If the columns of a matrix Q make up an orthonormal basis in L, then $P_L = Q\,Q^*$.*

Proof.
$$(1)\ (Q\,Q^*)^2 = Q\,(Q^*\,Q)\,Q^* = Q\,Q^*;$$
$$(2)\ (Q\,Q^*)^* = (Q^*)^*\,Q^* = Q\,Q^*. \quad \Box$$

9.6 Distances and orthoprojectors

Lemma 9.6.1 *Suppose orthoprojectors P_L and P_M correspond to subspaces L and M. Then*
$$\rho(L, M) = \|(I - P_M)\,P_L\|_2.$$

Proof.

$$\begin{aligned}
\rho(L, M) &= \|P_L\,x - P_M\,P_L\,x\|_2 \quad (\text{for some } x = P_L\,x,\ \|x\|_2 = 1) \\
&\leq \|P_L - P_M\,P_L\|_2 = \|(I - P_M)\,P_L\|_2.
\end{aligned}$$

Now, we show that $\|(I - P_M)\,P_L\|_2 \leq \rho(L, M)$. Let

$$\|(I - P_M)\,P_L\|_2 = \|(I - P_M)\,P_L\,y\|_2,\ \|y\|_2 = 1.$$

If $P_L \, y = 0$, then that is trivial. If $P_L \, y \neq 0$, then

$$\|(I - P_M) \, P_L \, y\|_2 \;\leq\; \left\| \frac{P_L \, y}{\|P_L \, y\|_2} - P_M \, \frac{P_L \, y}{\|P_L \, y\|_2} \right\|_2 \|P_L \, y\|_2 \;\leq\; \rho \, (L, \, M). \; \square$$

Theorem 9.6.1 *Suppose orthoprojectors P_L and P_M correspond to subspaces L and M. Then*

$$\mathrm{dist} \, (L, \, M) = \|P_L - P_M\|_2.$$

Proof. From Lemma 9.6.1,

$$\rho \, (L, \, M) = \|(I - P_M) \, P_L\|_2 = \|(P_L - P_M) \, P_L\|_2 \leq \|P_L - P_M\|_2.$$

$$\Rightarrow \quad \mathrm{dist} \, (L, \, M) \;\leq\; \|P_L - P_M\|_2.$$

Further, $\exists \, x \in \mathbb{C}^n, \; \|x\|_2 = 1 : \quad \|P_L - P_M\|_2 \;=\; \|(P_L - P_M) \, x\|_2$. By the Pythagor theorem,

$$
\begin{aligned}
\|(P_L \;-\; & P_M) \, x\|_2^2 \\
&= \; \|P_L \, (P_L - P_M) \, x\|_2^2 + \|(I - P_L) \, (P_L - P_M) \, x\|_2^2 \\
&= \; \|P_L \, (I - P_M) \, x\|_2^2 + \|(I - P_L) \, P_M \, x\|_2^2 \\
&\leq \; \|P_L \, (I - P_M)\|_2^2 \, \|(I - P_M) \, x\|_2^2 + \|(I - P_L) \, P_M\|_2^2 \, \|P_M \, x\|_2^2 \\
&\leq \; \max \, \{\|(P_L \, (I - P_M)\|_2^2, \; \|(I - P_L) \, P_M\|_2^2\}.
\end{aligned}
$$

It remains to note that

$$\|P_L \, (I - P_M)\|_2 = \|(P_L \, (I - P_M))^*\|_2 = \|(I - P_M) \, P_L\|_2 \;\leq\; \rho \, (L, \, M). \quad \square$$

Corollary 9.6.1 *The distance between subspaces is equal to the distance between their orthogonal complements.*

9.7   Subspaces of equal dimension

Lemma 9.7.1 *If a matrix*

$$Q = \begin{bmatrix} Q_{11} & Q_{12} \\ Q_{21} & Q_{22} \end{bmatrix}, \quad Q_{11} \in \mathbb{C}^{m \times m}, \; Q_{22} \in \mathbb{C}^{k \times k},$$

is unitary, then $\|Q_{12}\|_2 = \|Q_{21}\|_2$.

Proof. Consider the singular value decomposition $Q_{11} = U_1 \, \Sigma_1 \, V_1^*$ for the first diagonal block, and, then, take up another unitary matrix

$$\hat{Q} \;=\; \begin{bmatrix} U_1^* & 0 \\ 0 & I \end{bmatrix} Q \begin{bmatrix} V_1 & 0 \\ 0 & I \end{bmatrix} = \begin{bmatrix} \Sigma_1 & \hat{Q}_{12} \\ \hat{Q}_{21} & \hat{Q}_{22} \end{bmatrix}.$$

It is clear that $\|Q_{12}\|_2 = \|\hat{Q}_{12}\|_2$, $\|Q_{21}\|_2 = \|\hat{Q}_{21}\|_2$. At the same time,

$$
\begin{aligned}
\hat{Q} \, \hat{Q}^* = I \quad &\Rightarrow \quad \hat{Q}_{12} \, \hat{Q}_{12}^* = I - \Sigma_1^2, \\
\hat{Q}^* \, \hat{Q} = I \quad &\Rightarrow \quad \hat{Q}_{21}^* \, \hat{Q}_{21} = I - \Sigma_1^2. \quad \square
\end{aligned}
$$

Theorem 9.7.1 *Suppose that subspaces L and M are of equal dimension and unitary matrices $U = [U_1 \ U_2]$ and $V = [V_1 \ V_2]$ are such that*

$$\operatorname{im} U_1 = L, \quad \operatorname{im} V_1 = M.$$

Then

$$\operatorname{dist}(L, M) = \|U_1^* V_2\|_2 = \|U_2^* V_1\|_2.$$

Proof.

$$U^* (U_1 U_1^* - V_1 V_1^*) V = \begin{bmatrix} 0 & U_1^* V_2 \\ -U_2^* V_1 & 0 \end{bmatrix}. \quad \square$$

Corollary 9.7.1 *If $\dim L = \dim M$, then $\rho(L, M) = \rho(M, L)$.*

Proof.

$$U^* (U_1 U_1^*) (I - (V_1 V_1^*)) V = \begin{bmatrix} 0 & U_1^* V_2 \\ 0 & 0 \end{bmatrix},$$

$$V^* (V_1 V_1^*) (I - (U_1 U_1^*)) U = \begin{bmatrix} 0 & V_1^* U_2 \\ 0 & 0 \end{bmatrix}. \quad \square$$

9.8 The CS decomposition

The distance between subspaces is only one characteristic of their mutual disposition. More detailed information is given by the so-called principal angles between the two subspaces. We determine them from the CS decomposition of a unitary matrix coupled with the two subspaces.

Theorem 9.8.1 *Assume that Q is a unitary matrix of order n. Then, for any $m \le n/2$, there exist unitary matrices U_1, V_1 of order m and unitary matrices U_2, V_2 of order $n - m$ such that*

$$\begin{bmatrix} U_1 & 0 \\ 0 & U_2 \end{bmatrix} Q \begin{bmatrix} V_1 & 0 \\ 0 & V_2 \end{bmatrix} = \begin{bmatrix} C & S & 0 \\ -S & C & 0 \\ 0 & 0 & I_{n-2m} \end{bmatrix}, \qquad (*)$$

$$C = \operatorname{diag}(c_1, \dots, c_m), \quad S = \operatorname{diag}(s_1, \dots, s_m),$$

$$c_1 \ge \dots \ge c_m \ge 0,$$

$$0 \le s_1 \le \dots \le s_m,$$

$$c_i^2 + s_i^2 = 1, \quad i = 1, \dots, m.$$

The decomposition $(*)$ is said to be the CS decomposition of the unitary matrix Q.

We embark on the proof by setting Q in block form

$$\begin{bmatrix} Q_{11} & Q_{12} \\ Q_{21} & Q_{22} \end{bmatrix}, \quad Q_{11} \in \mathbb{C}^{m \times m},$$

and, then, by setting the singular value decomposition for the block $Q_{11} = U C V^*$. We obtain

$$\begin{bmatrix} U^* & 0 \\ 0 & I_{n-m} \end{bmatrix} Q \begin{bmatrix} V & 0 \\ 0 & I_{n-m} \end{bmatrix} = \begin{bmatrix} C & W_{12} \\ W_{21} & W_{22} \end{bmatrix} \equiv W.$$

The equations $W W^* = W^* W = I$ imply that

$$W_{12} W_{12}^* = W_{21}^* W_{21} = I_m - C^2 \equiv S^2,$$

and we see that there exist unitary matrices X, Y of order $n - m$, providing that

$$W_{12} Y = [S, \, 0], \quad X W_{21} = [-S, \, 0]^T.$$

(Here, we have some room for choice. For example, we could swap S and $-S$ or multiply them by arbitrary diagonal unitary matrices.) Complete the proof!

Corollary. *Let subspaces L and M be of equal dimension $m \le \frac{n}{2}$. Then there exist orthonormal bases $u_1, \ldots, u_m \in L$ and $v_1, \ldots, v_m \in M$ such that*

$$u_i^* v_j = \begin{cases} c_i, & i = j, \\ 0, & i \ne j, \end{cases}$$

where the quantities $c_1 \ge \ldots \ge c_m \ge 0$ are determined uniquely.

To prove the corollary, it is sufficient to regard the CS decomposition of the unitary matrix $U^* V$, where $U = [U_1 \ U_2]$, $V = [V_1 \ V_2]$ are unitary matrices such that $L = \operatorname{im} U_1$, $M = \operatorname{im} V_1$.

One may write

$$c_i = \cos \phi_i, \quad 0 \le \phi \le \frac{\pi}{2}.$$

The angles ϕ_i are referred to as the *principal angles* between the subspaces L and M.

9.9 Convergence of subspace iterations for the block diagonal matrix

First, we consider the subspace iterations for a block diagonal matrix. Let

$$A = \begin{bmatrix} A_1 & 0 \\ 0 & A_2 \end{bmatrix}, \quad A_1 \in \mathbb{C}^{m \times m}, \ A_2 \in \mathbb{C}^{r \times r}, \quad \exists \, A_1^{-1}. \qquad (*)$$

Denote by M the m-dimensional subspace spanned on the first m columns of the unity matrix. Obviously, M is A-invariant. To "cause" the method of subspace iterations to converge to M, we shall require that the initial m-dimensional subspace L be not "too far" from M:

$$\beta \equiv \rho(L, M) < 1. \qquad (**)$$

Lemma 9.9.1 *Given* (∗) *and* (∗∗), *the following inequality holds:*

$$\rho(AL, M) \leq \frac{\beta}{\sqrt{1 - \beta^2}} \, \|A_2\|_2 \, \|A_1^{-1}\|_2. \qquad (9.9.1)$$

Proof. Let $y = A\,x$, $y = \begin{bmatrix} y_1 \\ y_2 \end{bmatrix}$, $x = \begin{bmatrix} x_1 \\ x_2 \end{bmatrix}$, $x_1, y_1 \in \mathbb{C}^m$. Assume that $x \in L$, $\rho(AL, M) = \rho(y, M)$, $\|y\|_2 = 1$. Then

$$\rho\left(\frac{x}{\|x\|_2}, M\right) = \frac{\|x_2\|_2}{\|x\|_2} \leq \beta \quad \Rightarrow \quad \|x_2\|_2 \leq \frac{\beta}{\sqrt{1 - \beta^2}} \, \|x_1\|_2 \quad \Rightarrow$$

$$
\begin{aligned}
\rho(y, M) = \|y_2\|_2 &= \|A_2 \, x_2\|_2 \leq \|A_2\|_2 \, \|x_2\| \\
&\leq \|A_2\|_2 \, \frac{\beta}{\sqrt{1 - \beta^2}} \, \|x_1\|_2 \\
&\leq \|A_2\|_2 \, \frac{\beta}{\sqrt{1 - \beta^2}} \, \|A_1^{-1} \, y_1\|_2 \\
&\leq \|A_2\|_2 \, \frac{\beta}{\sqrt{1 - \beta^2}} \, \|A_1^{-1}\|_2. \quad \square
\end{aligned}
$$

Denote by λ_{m+1} the maximal in modulus eigenvalue for the block A_2, by λ_m the minimal in modulus eigenvalue for the block A_1, and set

$$\gamma = \gamma(A_1, A_2) \equiv |\lambda_{m+1}| \, / \, |\lambda_m|. \qquad (9.9.2)$$

Corollary 9.9.1 *If* $\gamma < 1$, *then* $\forall q \in (\gamma, 1) \, \exists \, c = c(q):$

$$\rho(A^k L, M) \leq c\,q^k, \quad k = 1, 2, \ldots.$$

Proof. Applying Lemma 9.9.1 to the matrix A^k,

$$\rho(A^k L, M) \leq \frac{\beta}{\sqrt{1 - \beta^2}} \, \|A_2^k\|_2 \, \|(A_1^{-1})^k\|_2.$$

We know that, for any matrix A, the sequence $\left(\|A^k\|_2\right)^{\frac{1}{k}}$ tends to the spectral radius of A (the maximal in modulus eigenvalue), as $k \to \infty$. Hence, $\forall \, \delta > 0 \, \exists \, k_\delta$ such that, for any $k > k_\delta$,

$$\|A_2^k\|_2 \, \|(A_1^{-1})^k\|_2 \leq (|\lambda_{m+1}| + \delta)^k \left(\frac{1}{|\lambda_m|} + \delta\right)^k.$$

For any $q > \gamma$, for all sufficiently small $\delta > 0$ the right-hand side of the above inequality will be less than q^k. $\square$

9.10 Convergence of subspace iterations in the general case

Assume that

$$A = Z \Lambda Z^{-1}, \quad \Lambda = \begin{bmatrix} A_1 & 0 \\ 0 & A_2 \end{bmatrix}, \qquad (9.10.1)$$

$$A_1 \in \mathbb{C}^{m \times m}, \quad A_2 \in \mathbb{C}^{r \times r}, \quad \exists A_1^{-1}.$$

First, consider a simple auxiliary lemma.

Lemma 9.10.1 *For any nonsingular matrix Z and for any subspaces $\mathcal{L}$, $\mathcal{M}$,*

$$\rho(Z \mathcal{L}, Z \mathcal{M}) \leq \text{cond}_2 Z \, \rho(\mathcal{L}, \mathcal{M}).$$

Proof. Let $x \in \mathcal{L}$, $\|x\|_2 = 1$, and assume that

$$\rho(Z \mathcal{L}, Z \mathcal{M}) = \rho\left(\frac{Z x}{\|Z x\|_2}, Z \mathcal{M}\right).$$

We find that

$$
\begin{aligned}
\rho\left(\frac{Z x}{\|Z x\|_2}, Z \mathcal{M}\right) &= \rho\left(\frac{Z x}{\|Z x\|_2}, \frac{Z}{\|Z x\|_2} \mathcal{M}\right) \\
&\leq \frac{1}{\|Z x\|_2} \|Z (x - z)\|_2 \quad (\forall z \in \mathcal{M}) \\
&\leq \frac{\|Z\|_2}{\|Z x\|_2} \|x - z\|_2 \\
&\leq \|Z\|_2 \|Z^{-1}\|_2 \, \rho(\mathcal{L}, \mathcal{M}). \quad \square
\end{aligned}
$$

Theorem 9.10.1 *Assume that A is of the form (9.10.1), M stands for the linear subspace spanned on the first m columns of the matrix Z, and an m-dimensional space L is such that*

$$\beta \equiv \rho(Z^{-1} L, Z^{-1} M) < 1.$$

Let γ be defined by (9.9.2). If $\gamma < 1$, then $\forall q \in (\gamma, 1) \, \exists \, c = c(q)$ such that

$$\rho(A^k L, M) \leq c q^k, \quad k = 1, 2, \ldots.$$

Proof. The general case reduces to that of the block diagonal matrix Λ for which the subspace $Z^{-1} M$ is invariant. (Why?) We already know that, for Λ, the iterations of the initial subspace $Z^{-1} L$ will converge to $Z^{-1} M$. Hence, for any $q \in (\gamma, 1)$,

$$\rho(\Lambda^k (Z^{-1} L), (Z^{-1} M)) \leq c q^k$$

Since $A^k = Z \Lambda^k Z^{-1}$, using Lemma 9.10.1, we find that

$$\rho(A^k L, M) = \rho(Z (\Lambda^k Z^{-1} L), Z (Z^{-1} M)) \leq \text{cond}_2 Z \, c \, q^k. \quad \square$$

If you have felt that the analysis above was altogether clear and natural, please take into account that it has taken some effort to achieve that clarity, mostly due to a nice recent work of L. Elsner and D. S. Watkins.[1]

Exercises

1. Suppose that H is an upper Hessenberg matrix of order n with a nonzero subdiagonal. Consider an upper triangular matrix

$$
T(\lambda) = \begin{bmatrix} 1 & & & & \\ 0 & & & & \\ \cdots & & H - \lambda I & & \\ 0 & & & & \\ 0 & 0 & \cdots & 0 & 1 \end{bmatrix}
$$

and the following system of equations:

$$
T(\lambda)\, [\phi_0(\lambda), \ldots, \phi_n(\lambda)]^T = [0, \ldots, 0,\ 1]^T,
$$

with respect to polynomials $\phi_0(\lambda), \ldots, \phi_n(\lambda)$. Prove that $\phi_0(\lambda)$ may differ only by a nonzero coefficient from the characteristic polynomial of H.

2. Devise an algorithm that computes the coefficients of the characteristic polynomial for an arbitrary Hessenberg matrix of order n in $\mathcal{O}(n^3)$ arithmetic operations.

3. Devise an algorithm that computes the coefficients of the characteristic polynomial of an arbitrary $n \times n$ matrix in $\mathcal{O}(n^3)$ arithmetic operations.

4. Prove that $\text{dist}(L, M) = 1$ if and only if L contains a nonzero vector orthogonal to M.

5. Prove that the distance between any two subspaces in $\mathbb{C}^n$ is equal to the distance between their orthogonal complements.

6. Expound a complete proof of the theorem about the CS decomposition of a unitary matrix.

7. A unitary matrix has the following block form:

$$
Q = \begin{bmatrix} Q_{11} & Q_{12} \\ Q_{21} & Q_{22} \end{bmatrix}, \quad Q_{11}, Q_{22} \in \mathbb{C}^{m \times m}.
$$

Prove that $|\det(Q_{12})| = |\det(Q_{21})|$.

---

[1]D.S.Watkins and L.Elsner. Convergence of algorithms of decomposition type for the eigenvalue problem. *Linear Algebra Appl.* 143: 19–47 (1991).

8. Let A be a nonsingular matrix of order n, and let L and M be subspaces in $\mathbb{C}^n$. Prove that

$$\text{dist}\,(A\,L,\ A\,M) \leq \text{cond}_2\,(A)\ \text{dist}\,(L,\ M).$$

9. If subspaces L and M are of equal dimension, then, for any orthonormal basis of L (the columns of U), there exists an orthonormal basis of M (the columns of V) such that

$$\|U - V\|_2 \leq \sqrt{2}\ \text{dist}\,(L,\ M).$$

10. Is it true that if $\rho\,(L,\ M) < 1$, $\ L,\ M \in \mathbb{C}^n$, then $\rho\,(Z\,L,\ Z\,M) < 1$ holds for any nonsingular matrix $Z \in \mathbb{C}^{n \times n}$?

Lecture 10

10.1 The QR algorithm

Given a dense unstructured matrix A of several hundred (thousand) order with all the eigenvalues (eigenvectors) wanted, a vehicle of choice is definitely the QR algorithm. Since the early 60s, it is used due to V. N. Kublanovskaya (Russia) and J. G. F. Francis (England). We begin with the

QR iterations in the orthodox form:

$$A_0 = A \in \mathbb{C}^{n \times n} \text{ (a given matrix)};$$
$$A_{k-1} = Q_k R_k \text{ (the } QR \text{ decomposition)}, and$$
$$A_k = R_k Q_k, \quad k = 1, 2, \ldots .$$

How do the eigenvalues come in? We merely reduce the problem for A to the same one for A_k (because A_k and A are similar) — in the hope that A_k becomes *simpler*. In effect, an $(n - m) \times m$ block in the lower-left corner of A_k tends to zero (under a certain hypothesis), and, sooner or later, we neglect "the zeroes", thus reducing the problem to that for the two blocks of order m and $n - m$ on the diagonal. But nobody applies the QR in the orthodox form. We shall see soon that, if $|\lambda_1| \geq \ldots |\lambda_m| > |\lambda_{m+1}| \geq \ldots \geq |\lambda_n|$, then, the entries of the $(n - m) \times m$ block diminish (next to always) up to a factor of $\gamma_m \equiv |\lambda_{m+1}|/|\lambda_m|$ at every step. *Convergence is slow* when $\gamma_m \approx 1$, and how to make γ_m smaller is probably to transfer to a "shifted" matrix $A - s I$. Eventually, we come to the

QR iterations with shifts:

$$A_0 = A;$$
$$A_{k-1} - s_k I = Q_k R_k \text{ (the } QR \text{ decomposition)},$$
$$A_k = R_k Q_k + s_k I, \quad k = 1, 2, \ldots .$$

10.2 Generalized QR algorithm

The QR iterations with shifts are a particular case of a more general algorithm involving polynomials f_k:

$$A_0 = A;$$
$$f_k(A_{k-1}) = Q_k R_k \text{ (the } QR \text{ decomposition)},$$
$$A_k = Q_k^{-1} A_{k-1} Q_k, \quad k = 1, 2, \dots .$$

The generalized QR algorithm is sometimes called the QR algorithm with *multishifts*. A multishift of degree $r = r(k)$ means a complete set of the roots of

$$f_k(x) = \prod_{i=1}^{r}(x - s_i^{(k)}).$$

One multishift of degree r is equivalent to a sequence of r multishifts of degree 1. (Explain why.)

10.3　Basic formulas

The QR analysis rests totally on the following formulas:

$$A_k = Z_k^{-1} A Z_k, \quad Z_k = Q_1 \dots Q_k; \tag{10.3.1}$$

$$p_k(A) \equiv \prod_{i=1}^{k} f_i(A) = Z_k U_k, \quad \text{and} \quad U_k = R_k \dots R_1. \tag{10.3.2}$$

Proof. The first is obvious. To prove the second (by induction), assume that we have already established the equality

$$f_2(A_1) \dots f_k(A_1) = (Q_2 \dots Q_k)(R_k \dots R_2).$$

Since $A_1 = Q_1^{-1} A Q_1$, we find that

$$
\begin{aligned}
f_2(Q_1^{-1} A Q_1) \dots f_k(Q_1^{-1} A Q_1) &= (Q_1^{-1} f_2(A) Q_1) \dots (Q_1^{-1} f_k(A) Q_1) \\
&= Q_1^{-1} f_2(A) \dots f_k(A) Q_1,
\end{aligned}
$$

which is followed by

$$Q_1^{-1} f_2(A) \dots f_k(A) Q_1 = (Q_2 \dots Q_k)(R_k \dots R_2).$$

Premultiply both sides by Q_1, and postmultiply by R_1. Recall that $f_1(A) = Q_1 R_1$, and so, arrive at (10.3.2).　□

10.4　The QR iteration lemma

Denote by $G(m, n, X)$ a set of all matrices $A \in \mathbb{C}^{n \times n}$ satisfying the following requirements:

(1) $A = X\Lambda X^{-1}$, $\quad \Lambda = \begin{bmatrix} \Lambda_1 & 0 \\ 0 & \Lambda_2 \end{bmatrix}$,

$\quad \Lambda_1 \in \mathbb{C}^{m \times m}$, $\Lambda_2 \in \mathbb{C}^{r \times r}$, $m + r = n$.

(2) The leading principal submatrix of order m in X^{-1} is nonsingular.

Lemma 10.4.1 *Consider one generalized QR iteration with a polynomial* $f(\lambda)$ *for a matrix* $A_0 \in G(m, n, X)$:

$$f(A_0) = QR, \quad A_1 = Q^* A_0 Q,$$

and set

$$A_0 = \begin{bmatrix} A_{11}^{(0)} & A_{12}^{(0)} \\ A_{21}^{(0)} & A_{22}^{(0)} \end{bmatrix}, \quad A_1 = \begin{bmatrix} A_{11}^{(1)} & A_{12}^{(1)} \\ A_{21}^{(1)} & A_{22}^{(1)} \end{bmatrix}, \quad A_{11}^{(0)}, A_{11}^{(1)} \in \mathbb{C}^{m \times m}.$$

Assume that blocks $F_1 \equiv f(\Lambda_1)$ *and* $F_2 \equiv f(\Lambda_2)$ *are nonsingular. Then*

$$||A_{21}^{(1)}||_2 \le c_1 (1 + c_2 \phi)^2 \, \phi \, ||A_{21}^{(0)}||_2,$$

where

$$\phi = ||F_2||_2 \, ||F_1^{-1}||_2$$

and the constants c_1, $c_2 > 0$ *depend only upon* m *and* X.

Proof. Owing to requirement (2) on the class $G(m, n, X)$, the matrix X^{-1} admits the following block LU decomposition:

$$X^{-1} = LU \equiv \begin{bmatrix} I_m & 0 \\ L_{21} & I_r \end{bmatrix} \begin{bmatrix} U_{11} & U_{12} \\ 0 & U_{22} \end{bmatrix}, \quad U_{11} \in \mathbb{C}^{m \times m}.$$

The matrices L and U are nonsingular (why?), and their inverses are of the same block structure:

$$L^{-1} = \begin{bmatrix} I_m & 0 \\ -L_{21} & I_r \end{bmatrix}, \quad U^{-1} = \begin{bmatrix} U_{11}^{-1} & -U_{11}^{-1} U_{12} U_{22}^{-1} \\ 0 & U_{22}^{-1} \end{bmatrix}.$$

In accordance with the lemma's hypotheses, the block diagonal matrix $F \equiv f(\Lambda)$ is nonsingular, and it is easy to verify that

$$A_1 = RU^{-1} F^{-1} \{ F \, L^{-1} \Lambda L \, F^{-1} \} FUR^{-1}.$$

The matrix $Q = X(FLF^{-1})(FUR^{-1})$ is unitary $\implies$

$$||FUR^{-1}||_2 \le ||X^{-1}||_2 \, ||FL^{-1}F^{-1}||_2.$$

The matrix Q^{-1} is unitary as well $\implies$

$$||RU^{-1}F^{-1}||_2 \le ||X||_2 \, ||FLF^{-1}||_2.$$

Then,

$$FLF^{-1} = \left[\begin{array}{cc} I & 0 \\ F_2 L_{21} F_1^{-1} & I \end{array} \right],$$

and hence,

$$\|FLF^{-1}\|_2 \leq \|X\|_2 \, (1 + \|L_{21}\|_2 \phi).$$

Analogously,

$$\|FL^{-1}F^{-1}\|_2 \leq \|X^{-1}\|_2 \, (1 + \|L_{21}\|_2 \phi).$$

Now, consider the block partitioning

$$F(L^{-1}\Lambda L)F^{-1} = \left[\begin{array}{cc} F_1 & 0 \\ F_2 \{L^{-1}\Lambda L\}_{21} F_1^{-1} & F_2 \end{array} \right].$$

Since $L^{-1}\Lambda L = U A U^{-1}$, we infer $\{L^{-1}\Lambda L\}_{21} = U_{22} A_{21}^{(0)} U_{11}^{-1}$, which entails

$$\|\{L^{-1}\Lambda L\}_{21}\|_2 \leq (\|U_{22}\|_2 \, \|U_{11}^{(-1)}\|_2) \, \|A_{21}^{(0)}\|_2.$$

Finally,

$$\begin{array}{rl} \|A_{21}^{(1)}\|_2 & \leq \left\| RU^{-1}F^{-1} \left[\begin{array}{cc} 0 & 0 \\ \Lambda_2 \{L^{-1}\Lambda L\}_{21} \Lambda_1^{-1} & 0 \end{array} \right] FUR^{-1} \right\|_2 \\[2mm] & \leq c_1 \, (1 + c_2\phi)^2 \|A_{21}^{(0)}\|_2, \end{array}$$

where

$$c_1 = \|X\|_2 \, \|X^{-1}\|_2 \, \|U_{22}\|_2 \, \|U_{11}^{-1}\|_2, \quad c_2 = \|L_{21}\|_2. \qquad \square$$

10.5   Convergence of the QR iterations

Theorem 10.5.1 *Suppose that a matrix A satisfies the following requirements:*

(1) $A = X \Lambda X^{-1}$, $\quad \Lambda = \left[\begin{array}{cc} \Lambda_1 & 0 \\ 0 & \Lambda_2 \end{array} \right]$, $\Lambda_1 \in \mathbb{C}^{m \times m}$, $\Lambda_2 \in \mathbb{C}^{r \times r}$.

(2) $|\lambda_1| \geq \ldots \geq |\lambda_m| > |\lambda_{m+1}| \geq \ldots \geq |\lambda_{m+r}| > 0$,
$\{\lambda_1, \ldots, \lambda_m\} = \lambda(\Lambda_1), \quad \{\lambda_{m+1}, \ldots, \lambda_{m+r}\} = \lambda(\Lambda_2).$

(3) *The leading principal submatrix of order m in X^{-1} in nonsingular.*

Then the orthodox QR iterations generate the matrices

$$A_k = \left[\begin{array}{cc} A_{11}^{(k)} & A_{12}^{(k)} \\ A_{21}^{(k)} & A_{22}^{(k)} \end{array} \right]$$

such that

$$A_{21}^{(k)} \to 0 \quad for \quad k \to 0.$$

Moreover, if $\gamma \equiv |\lambda_{m+1}|/|\lambda_m|$, then

$$\forall q \in (\gamma, 1) \ \exists c = c(q): \ \|A_{21}^{(k)}\|_2 \leq c \, q^k, \quad k = 1, 2, \ldots.$$

Proof. Note that k steps of the orthodox QR can be viewed as one step of the generalized QR with a polynomial $f(\lambda) = \lambda^k$:

$$A^k = Z_k U_k, \qquad A_k = Z_k^* A Z_k.$$

By the QR iteration lemma,

$$\|A_{21}^{(k)}\|_2 \ \le \ c \, \|\Lambda_2^k\|_2 \, \|\Lambda_1^{-k}\|_2,$$

and it remains to be observed that, for any $\delta > 0$, for sufficiently large k,

$$\|\Lambda_2^k\|_2 \le (|\lambda_{m+1}| + \delta)^k, \qquad \|\Lambda_1^{-k}\|_2 \le \left(\frac{1}{|\lambda_m|} + \delta\right)^k. \ \ \square$$

Corollary 1. *Under the hypothesis of this theorem, if Λ is a diagonal matrix, then, for some $c > 0$,*

$$\|A_{21}^{(k)}\|_2 \ \le \ c \left(\frac{|\lambda_{m+1}|}{|\lambda_m|}\right)^k, \quad k = 1, 2, \ldots.$$

Corollary 2. *Let*

$$A = X\Lambda X^{-1}, \quad \Lambda = \mathrm{diag}\,(\lambda_1, \ldots, \lambda_n), \qquad (10.5.1)$$

$$|\lambda_1| > \ldots > |\lambda_n| > 0. \qquad (10.5.2)$$

If the eigenvector matrix X^{-1} is strongly regular (i.e., all its leading principal submatrices are nonsingular), then

$$\lim_{k\to\infty} \{A_k\}_{ij} = 0, \quad i > j; \qquad \lim_{k\to\infty} \mathrm{diag}\, A_k = \Lambda.$$

10.6 Pessimistic and optimistic

If the requirements considered above are not fulfilled, then there might be no subdiagonal blocks that converge to zero.

A pessimistic remark: in general, the orthodox QR iterations are not bound to converge. (Produce an example.)

An optimistic remark: the requirements (1), (2), and (3) in Section 10.5, can always be met through arbitrarily small perturbations. (Why?)

In the convergence proof above, we developed somewhat an elegant elementary approach of Wilkinson to the QR. Among other requirements, that of strong regularity of X^{-1} might not seem very natural. It is still essential as an attribute of a proof where the LU decomposition of X^{-1} features in. All the same, *ought* it to be essential?

With the optimistic remark in mind, one might decide not to waste efforts. I still think that those who feel unsatisfied would be interested in becoming aware how to do without the strong regularity requirement. To this end, we need *the Bruhat decomposition* instead of the *LU*.

10.7   Bruhat decomposition

The *basic Bruhat decomposition* of a nonsingular matrix A is defined as

$$A = L_1 \, \Pi \, L_2,$$

where $\Pi = \Pi(A)$ is a permutational matrix and L_1 and L_2 are nonsingular lower triangular matrices. The *modified Bruhat decomposition* is of the form

$$A = LPU,$$

where $P = P(A)$ is a permutational matrix, and L and U are nonsingular lower and upper triangular matrices, respectively.

Theorem 10.7.1 *The modified and basic Bruhat decompositions exist for any nonsingular matrix A. The permutational matrices $\Pi = \Pi(A)$ and $P = P(A)$ are defined uniquely and possess the properties*

$$\Pi(A) = P(A\,J)\,J, \qquad P(A) = \Pi(A\,J)\,J, \qquad \text{where} \quad J = \begin{bmatrix} & & 0 & & 1 \\ & & & \cdot^{\cdot^{\cdot}} & \\ & 1 & & & 0 \end{bmatrix}.$$

Proof. We prove the existence of the modified Bruhat decomposition by a construction. Show that A can be reduced to a permutational matrix by a sequence of pre- and postmultiplications using appropriate lower and upper triangular matrices.

Consider the first (from the left) nonzero entry in the first row of A. Call it the pivoting entry. Postmultiplication by an upper triangular matrix can kill all subsequent entries in the first row and make the pivoting entry equal to 1. Next, premultiplication by a lower triangular matrix can be used to zero all entries which are located below the pivoting one in its column. Once this is done, we find a new pivoting entry, that is, the first nonzero entry in the second row of the current matrix. Using postmultiplication, we annihilate all entries to the right of the pivoting one in the second row, and using premultiplication, then, we get rid of all entries below the pivoting one in its column, and so on. In the end, we arrive at some permutational matrix P.

Consider the sub-rows

$$r_i(j) \equiv [a_{i1}, \dots, a_{ij}],$$

and denote by $\sigma(i;\,A)$ the maximal index among those j for which $r_i(j) \notin$ span $\{r_1(j), \dots, r_{i-1}(j)\}$. It is easy to verify that

$$\sigma(i;\,A) = \sigma(i;\,LAU)$$

for any nonsingular lower and upper triangular matrices L and U. Since the unit in the column i of P is located exactly at the position $j = \sigma(i;\,P) = \sigma(i,\,A)$, the permutational matrix P is determined uniquely.

It remains to ascertain the following relationships:

$$AJ = LPU \quad \Leftrightarrow \quad A = L\,(PJ)\,(JUJ)$$
$$AJ = L_1 \Pi L_2 \quad \Leftrightarrow \quad A = L_1\,(\Pi J)\,(JL_2 J). \qquad \square$$

10.8 What if the matrix X^{-1} is not strongly regular

Theorem 10.8.1 *Suppose that A satisfies (10.5.1) and (10.5.2). The orthodox QR algorithm generates the matrices A_k such that*

$$\{A_k\}_{ij} \to 0 \quad for \quad i > j, \quad and \qquad (10.8.1)$$
$$\operatorname{diag}(A_k) \to \operatorname{diag}(P^{-1}\Lambda P), \qquad (10.8.2)$$

where P is the permutational matrix from the modified Bruhat decomposition of $X^{-1} = LPU$.

Proof. Using the basic formulas for the orthodox QR case we obtain

$$A_k = (U_k U^{-1} P^{-1} \Lambda^{-k} P) \{P^{-1}\Lambda^k [L^{-1}\Lambda L] \Lambda^{-k}P\} (P^{-1}\Lambda^k PUU_k^{-1}).$$

The matrices in the round brackets are mutually inverse upper triangular matrices. They are bounded uniformly in k (prove this). On the strength of (10.5.2) and due to an upper triangular form of $L^{-1}\Lambda L$,

$$\Lambda^k [L^{-1}\Lambda L] \Lambda^{-k} \to \operatorname{diag}(L^{-1}\Lambda L) = \Lambda.$$

Consequently, the matrix in curved brackets is of the form $P^{-1}\Lambda P + F_k$, where $F_k \to 0$. We finish by taking into account that

$$\operatorname{diag}(A_k) = P^{-1}\Lambda P + (U_k U^{-1} P^{-1} \Lambda^{-k}) F_k (P^{-1}\Lambda^k PUU_k^{-1}). \quad \square$$

To sum up, if the eigenvector matrix X^{-1} is not strongly regular, then the diagonal entries of A_k (for all sufficiently large k) approximate the eigenvalues of A taken in another order (not with respect to a decrease of moduli). I still believe that watching this in practice is next to impossible.

10.9 The QR iterations and the subspace iterations

Write $Z_k = [z_1^{(k)}, \ldots, z_n^{(k)}]$, and consider the subspaces

$$L_m^k \equiv \operatorname{span} \{z_1^{(k)}, \ldots, z_m^{(k)}\}.$$

Let the subspace $L_m \equiv L_m^0$ be spanned on the first m columns of the unity matrix.

Assume that A is nonsingular. Then

$$L_m^k = A^k L_m, \quad m = 1, \ldots, n.$$

Therefore, one QR iteration generates (virtually) n subspaces $L_1^k, \ldots, L_n^k$, which would arise at the kth step of the subspace iterations with the matrix A and initial subspaces $L_1, \ldots, L_n$.

We are already aware of what can provide for convergence of the subspace iterations to some invariant subspace. In particular, let A be diagonalizable and all its eigenvalues different in modulus:

$$A = X \operatorname{diag}(\lambda_1, \ldots, \lambda_n) X^{-1}, \quad |\lambda_1| > \ldots > |\lambda_n|.$$

Let

$$\rho\left(X^{-1} L_m, \ X^{-1} M_m\right) < 1 \quad \forall m, \tag{10.9.1}$$

where M_m is an A-invariant subspace spanned on the first m columns of X. Then

$$\forall m \quad L_m^k \ \to \ M_m \text{ for } k \ \to \ \infty.$$

If the subspace L_m^k is A-invariant, then the matrix $A_k = Z_k^* A Z_k$ possesses a zero subdiagonal block. It is intuitively clear that the closer the subspace to an A-invariant subspace, the smaller should be that subdiagonal block. Here is how we can put it quantitatively:

Lemma 10.9.1 *Assume that*

$$T = \begin{bmatrix} T_{11} & T_{12} \\ T_{21} & T_{22} \end{bmatrix} \in \mathbb{C}^{n \times n}, \quad T_{11} \in \mathbb{C}^{m \times m}, \ T_{22} \in \mathbb{C}^{r \times r}.$$

Let L be spanned on the first m columns of the unity matrix, and let M be an arbitrary T-invariant subspace of dimension m in $\mathbb{C}^n$. Then

$$\|T_{21}\|_2 \ \leq \ 3\,\rho(L, M)\,\|T\|_2.$$

Proof. Consider an orthonormal basis in M, and enlarge it to be an orthonormal basis in $\mathbb{C}^n$. From the vectors obtained, we make up the unitary matrix

$$U = \begin{bmatrix} U_{11} & U_{12} \\ U_{21} & U_{22} \end{bmatrix}, \quad M = \operatorname{im} \begin{bmatrix} U_{11} \\ U_{21} \end{bmatrix}.$$

Since M is invariant with respect to T,

$$\begin{bmatrix} T_{11} & T_{12} \\ T_{21} & T_{22} \end{bmatrix} \begin{bmatrix} U_{11} & U_{12} \\ U_{21} & U_{22} \end{bmatrix} = \begin{bmatrix} U_{11} & U_{12} \\ U_{21} & U_{22} \end{bmatrix} \begin{bmatrix} R_{11} & R_{12} \\ 0 & R_{22} \end{bmatrix}$$

for some blocks R_{11}, R_{12}, R_{22}. Hence,

$$T_{21} = [U_{21} \ U_{22}] \begin{bmatrix} R_{11} & R_{12} \\ 0 & R_{22} \end{bmatrix} \begin{bmatrix} U_{11}^* \\ U_{12}^* \end{bmatrix}.$$

Thus, we arrive at $\|T_{21}\|_2 \leq 3\|U_{21}\|_2 \|T\|_2$. What is left is to note that $\rho(M, L) = \|U_{21}\|_2 = \|U_{21} y\|_2$ for some y of unit length. (Why?) □

We have one more proof of the fact that, under rather general assumptions, all the subdiagonal entries of A_k must tend to zero as $k \ \to \ \infty \ \Rightarrow \ $ the diagonal entries of A_k approximate the eigenvalues of A.

Note that the requirement (10.9.1) is equivalent to that of the strong regularity of X^{-1} (prove this).

Exercises

1. Produce an example of a matrix for which the orthodox QR iterations do not converge.

2. Assume that an upper triangular matrix $A \in \mathbb{C}^{n \times n}$ possesses pairwise distinct eigenvalues $\lambda_1, \ldots, \lambda_n$ and a sequence of matrices A_k is such that

$$\{A_k\}_{ij} \to \{A\}_{ij} \quad \text{for} \quad i \geq j.$$

Prove that, for any k, the eigenvalues of $\lambda_i(A_k)$ can be listed so that

$$\lambda_i(A_k) \to \lambda_i, \quad i = 1, \ldots, n.$$

3. Prove that if $\|L\|_2 \leq \|M\|_2$, then

$$\left\| \begin{bmatrix} I & 0 \\ L & I \end{bmatrix} \right\|_2 \leq \left\| \begin{bmatrix} I & 0 \\ M & I \end{bmatrix} \right\|_2 .$$

4. Prove that the requirement (10.9.1) is equivalent to the strong regularity of the matrix X^{-1}.

Lecture 11

11.1 Quadratic convergence

Suppose that the generalized QR algorithm exploits polynomials f_k (multishifts) of the same degree r on every step.

A multishift is called a *Rayleigh multishift* in the case where f_k is chosen to be the characteristic polynomial of the $r \times r$ block $A_{22}^{(k)}$:

$$A_k = \begin{bmatrix} A_{11}^{(k)} & A_{12}^{(k)} \\ A_{21}^{(k)} & A_{22}^{(k)} \end{bmatrix}, \quad A_{11}^{(k)} \in \mathbb{C}^{m \times m}, \ A_{22}^{(k)} \in \mathbb{C}^{r \times r}.$$

Let $G'(m, n, X)$ include those and only those matrices from $G(m, n, X)$ of which the blocks Λ_1 and Λ_2 are diagonal:

$$\Lambda_1 = \text{diag}\{\lambda_1, \ldots, \lambda_m\}, \quad \Lambda_2 = \text{diag}\{\lambda_{m+1}, \ldots, \lambda_{m+r}\}, \quad m + r = n;$$

$$|\lambda_1| \geq \ldots \geq |\lambda_m| > |\lambda_{m+1}| \geq \ldots \geq |\lambda_{m+r}|.$$

Theorem 11.1.1 *Let $A \in G'(m, n, X)$, and assume that the generalized QR algorithm with the Rayleigh multishifts of degree r is convergent, i.e.,*

$$\varepsilon_k \equiv \|A_{21}^{(k)}\|_2 \to 0.$$

Then it converges quadratically: $\exists \ \delta, \ c > 0 : \varepsilon_k \leq \delta \Rightarrow \varepsilon_{k+1} \leq c \, \varepsilon_k^2$.

Proof. Consider one step of the generalized QR algorithm:

$$f_k(A_{k-1}) = QR, \quad A_k = Q^* RQ,$$

and apply the QR iteration lemma:

$$\varepsilon_k \leq c \, \varepsilon_{k-1} \, \alpha_k \, \beta_k, \quad \text{where} \quad \alpha_k = \|f_k(\Lambda_2)\|_2, \quad \beta_k = \|(f_k(\Lambda_1))^{-1}\|_2.$$

Denote by $s_1, \ldots, s_r$ the eigenvalues of the block A_{22}^{k-1}. By the Bauer-Fike and the Gerschgorin theorems, for sufficiently small ε_{k-1}, we obtain

$$\min_{1 \leq i \leq r} |\lambda_{m+j} - s_i| \leq \text{cond}_2(X) \, \varepsilon_{k-1} \quad j = 1, \ldots, r.$$

Therefore,

$$\left| f_k(\lambda_{m+j}) \right| = \left| \prod_{i=1}^{r} (\lambda_{m+j} - s_i) \right| \leq c_1 \, \varepsilon_{k-1}, \quad j = 1, \ldots, r,$$

and

$$\left| f_k(\lambda_j) \right| = \left| \prod_{i=1}^{r} (\lambda_j - s_i) \right| \geq c_2 \, > \, 0, \quad j = 1, \ldots, m,$$

where $c_1, c_2 > 0$ do not depend on k (prove this). It follows that

$$\alpha_k \leq c_1 \, \varepsilon_{k-1}, \qquad \beta_k \geq c_2 \, > \, 0. \quad \square$$

Note that, prior to proving the quadratic convergence, we *assume* that a convergence takes place. There has been no theorem yet on the global convergence of the QR algorithm with the Rayleigh multishifts.

11.2   Cubic convergence

Theorem 11.2.1 *Let $A \in G'(m, n, X)$, and assume that the generalized QR algorithm with Rayleigh multishifts of order r is convergent, i.e.,*

$$\varepsilon_k \equiv \| A_{21}^{(k)} \|_2 \, \to \, 0,$$

and, moreover, for all k,

$$c_1 \, \| A_{21}^{(k)} \|_2 \leq \| A_{12}^{(k)} \|_2 \leq c_2 \, \| A_{21}^{(k)} \|_2, \tag{11.2.1}$$

where $c_1, c_2 > 0$ do not depend on k. Then it converges cubically:

$$\exists \, \delta, c \, > \, 0 : \varepsilon_k \leq \delta \, \Rightarrow \, \varepsilon_{k+1} \leq c \, \varepsilon_k^3.$$

From Theorem 11.1.1, we know that the convergence is at least quadratic. Let us follow the same logic and notation. A novelty that emanates from the condition (11.2.1) is the bound

$$\alpha_k \leq c_1 \varepsilon_{k-1}^2.$$

The proof can be based on the following analog of the Bauer–Fike theorem.

Theorem 11.2.2 (Elsner–Watkins) *Let*

$$A = \begin{bmatrix} A_{11} & 0 \\ 0 & A_{22} \end{bmatrix}, \quad A_{11} = X_1 \Lambda_1 X_1^{-1}, \; A_{22} = X_2 \Lambda_2 X_2^{-1},$$

$$\Lambda_1 = \operatorname{diag}(\lambda_1, \ldots, \lambda_m), \quad \Lambda_2 = \operatorname{diag}(\lambda_{m+1}, \ldots, \lambda_{m+r}).$$

If μ is an eigenvalue of $A + F$, where $F = \begin{bmatrix} 0 & F_{12} \\ F_{21} & 0 \end{bmatrix}$, then

$$\min_{1 \leq i \leq m} |\mu - \lambda_i| \, \min_{1 \leq j \leq r} |\mu - \lambda_{m+j}| \, \leq \, \operatorname{cond}_2(X_1) \operatorname{cond}_2(X_2) \, \|F_{12}\|_2 \, \|F_{21}\|.$$

Proof. If $\mu \in \lambda(A)$, then the inequality is trivial. Let $\mu \notin \lambda(A)$. Consider the following matrices:

$$M_1 = \left[\begin{array}{cc} \Lambda_1 - \mu I & X_1 F_{12} X_2^{-1} \\ X_2 F_{21} X_1^{-1} & \Lambda_2 - \mu I \end{array} \right],$$

$$M_2 = \left[\begin{array}{cc} \Lambda_1 - \mu I & X_1 F_{12} X_2^{-1} \\ 0 & (\Lambda_2 - \mu I) - X_2 F_{21} X_1^{-1} (\Lambda_1 - \mu I)^{-1} X_1 F_{12} X_2^{-1} \end{array} \right],$$

and

$$M_3 = (I - (\Lambda_2 - \mu I)^{-1} X_2 F_{21} X_1^{-1} (\Lambda_1 - \mu I)^{-1} X_1 F_{12} X_2^{-1}).$$

Clearly, M_1 is singular $\Rightarrow$ M_2 is the same $\Rightarrow$ M_3 is also singular.

$$\Rightarrow \qquad \|(\Lambda_2 - \mu I)^{-1} X_2 F_{21} X_1^{-1} (\Lambda_1 - \mu I)^{-1} X_1 F_{12} X_2^{-1}\|_2 \geq 1. \quad \square$$

Corollary 11.2.1 *If A is Hermitan, then, under hypotheses of Theorem 11.2.1, the generalized QR algorithm with Rayleigh multishifts converges cubically.*

To prove this, it is sufficient to observe that A_k will also be Hermitian, which provides (11.2.1) with $c_1 = c_2$.

11.3 What makes the QR algorithm efficient

Shifts and multishifts are indispensable attributes of an efficient QR algorithm. But it is not enough.

The problem is that a single QR iteration for an unstructured matrix requires $\mathcal{O}(n^3)$ arithmetic operations. It is far too expensive, even if the number of iterations is modest (usually about 5 iteration per an eigenvalue).

Luckily, there is an utterly simple device to make the iteration really inexpensive. Remember that any matrix A can be transformed to a unitarily similar upper Hessenberg matrix H by means of reflections or rotations . Once this is done, the QR algorithm can be applied to the Hessenberg matrix $A_0 = H$.

The Hessenberg transformation itself requires, of course, the same $\mathcal{O}(n^3)$ operations. What we gain is that now any single QR iteration will be performed in $\mathcal{O}(n^2)$ operations!

We have a cut-rate iteration because of *invariance property of the Hessenberg form* with respect to QR iterations. In other words, if H is an upper Hessenberg matrix, then a QR iteration of the form

$$f(H) = QR, \quad H_1 = Q^{-1} HQ \qquad (11.3.1)$$

can be performed so that H_1 remains an upper Hessenberg matrix.

It is sufficient to prove this for a multishift of order 1. (Why?) In this case, the QR decomposition can be computed via rotations eliminating the entries along the lower subdiagonal: $Q^* = G_{n\,n-1} \ldots G_{21}$. That the matrix $H_1 = RQ = RG_{21}^* \ldots G_{n\,n-1}^*$ is upper Hessenberg can be checked straightforward (do this).

If A is Hermitian, H is tridiagonal. Hence, H_1 is also tridiagonal. (Why?) In this case, we implement a single QR iteration through $\mathcal{O}(n)$ operations!

11.4   Implicit QR iterations

If the roots for f_k are known, then the multishift can be carried out through a sequence of multishifts of order 1. This is not always desirable.

For example, in the case of real entries, the roots of f_k might be complex, which causes us to use complex arithmetic. On the other hand, in the case of real entries, the coefficients of the characteristic polynomial f_k defining the Rayleigh multishift are still real (prove it).

There might be an implicit implementation of the QR iteration that does not reduce it to a sequence of multishifts of order 1. On input, we have the coefficients of f_k (the characteristic polynomial for $A_{22}^{(k)}$), not the roots. The following observation lies behind the algorithm.

Lemma 11.4.1 *Let A be a Hessenberg matrix, and assume that two Hessenberg matrices B and C with nonzero subdiagonals are such that the equations*

$$B = P^*AP, \quad C = Q^*AQ$$

hold for some unitary matrices P and Q of which the first columns are collinear. Then a diagonal unitary matrix D exists that provides that

$$P = QD, \quad B = D^*CD.$$

Proof. By induction, if $p_i = q_i d_i$, $|d_i| = 1$, $i = 1, \ldots, k$, then $b_{ij} = d_i^* c_{ij} d_j$ for $1 \leq i, j \leq k$, and we proceed as follows:

$$
\begin{aligned}
p_{k+1} b_{k+1\,k} &= A p_k - \sum_{i=1}^{k} p_i b_{ik} \\
&= A q_k d_k - \sum_{i=1}^{k} q_i d_i (d_i^* c_{ik} d_k) \\
&= \left(A q_k - \sum_{i=1}^{k} q_i (d_i d_i^*) c_{ik} \right) d_k \\
&= q_{k+1} c_{k+1\,k} d_k. \quad \square
\end{aligned}
$$

At every step of the QR algorithm, we go from some Hessenberg matrix H to a unitarily similar Hessenberg matrix $H_1 = Q^*HQ$. In machine arithmetic, one may always regard the subdiagonal of H_1 as nonzero. However we come to H_1, it is sufficient to ensure that the first column of the corresponding matrix Q should be the same as it were if we would have performed the standard (explicit) QR step.

Implicit QR iteration:

(1) Find the first column h of the matrix $f_k(H)$.

(2) Find the reflection matrix V_0 such that $V_0^* h = [*, 0, \ldots, 0]^T$.

(3) Find the matrix $W_0 = V_0^* H V_0$.

(4) By reflections, reduce W_0 to an unitarily similar matrix $W_1 = V_1^* W_0 V_1$ (with V_1 being the product of $n - 1$ reflection matrices).

(5) Set $H_1 = W_1$.

It is easy to see that $H_1 = (V_0 V_1)^* H (V_0 V_1)$. The first column of $V_0 V_1$ coincides with that of V_0 (prove it). Simultaneously, the first column of V_0 is the same as the first column of Q in some QR decomposition of $f_k(H)$. (Why?)

11.5   Arrangement of computations

A "clever" routine for the QR algorithm starts with seeking a diagonal matrix D that makes the rows and columns of $A_D = D^{-1} A D$ as close in length as possible. Such a leveling procedure can improve the condition numbers for the eigenvalues dramatically (sometimes by several orders).

Then we reduce A_D to a Hessenberg form A_0, and proceed as follows:

(1) Scan the subdiagonal in A_0, and replace every sufficiently small entry (in accordance with some "smallness" criterion) by zero.

(2) Choose a nonempty diagonal block in A_0 located in between the nearest-to-bottom pair of two successive zeroes on the subdiagonal. If there is no nonempty block, then quit (if so, the matrix has acquired the triangular form).

(3) For the chosen block, perform a single QR iteration (with a shift), then repeat the actions from (1).

The arrangement of computations as above is not bound to manifest high performance on a parallel computer. From this point of view, the multi-shifts seem especially attractive. Preliminary experiments with an implicit implementation of multishifts showed somewhat unsatisfactory results (the roundoffs were suspected to slow down the convergence). However, in 1994 D. S. Watkins[1] discovered that implicit multishifts work when implemented

---

[1] D.S.Watkins. Shifting strategies for the parallel QR-algorithm. *SIAM J. Sci. Comput.* 15 (4): 953–958 (1994).

in the pipeline fashion as a chained sequence of implicit multishifts of smaller degrees. The procedure was found to be numerically stable and maintained fast convergence.

11.6 How to find the singular value decomposition

First of all, we multiply $A \in \mathbb{C}^{n \times n}$ from both sides by unitary matrices and obtain an upper bidiagonal matrix B. Without loss of generality, we dare claim that B is a real matrix. (Why?)

If B has a zero on the diagonal, then by multiplying B in both sides by unitary matrices, we can obtain a block matrix of the form $\begin{bmatrix} B_1 & 0 \\ 0 & B_2 \end{bmatrix}$ with bidiagonal blocks B_1 and B_2.

Indeed, let $b_{kk} = 0$. The entry $b_{k\,k+1}$ can be nullified by a left-side rotation of the rows k and $k + 1$. A nonzero that might arise in the position $(k, k + 2)$ can be annihilated by a left-side rotation of the rows k and $k + 2$. A nonzero that might appear in the position $(k, k + 3)$ can be eliminated by a left-side rotation of the rows k and $k+3$, and so on. In the end, we obtain the block B_2. In a similar manner, we can obtain B_1 using right-side rotations of columns.

Thus, without loss of generality, we find that B is a nonsingular matrix. Now, we may freely apply the QR algorithm with shifts to a real symmetric tridiagonal matrix $T \equiv B^T B$. As soon as we compute the decomposition $B^T B = V \Sigma^2 V^T$ with an orthogonal matrix V and a diagonal matrix $\Sigma > 0$, we set $U = BV\Sigma^{-1}$. The matrix U will be unitary (why?), and, obviously, $B = U\Sigma V^T$.

In 1965, G. Golub and V. Kahan discovered that the implicit form of the QR iterations allows one not to capture the tridiagonal matrices explicitly. Consider a single QR iteration for tridiagonal matrices: $T - sI = QR$, $T_1 = Q^T TQ$. Set the Cholesky decomposition for T_1 as $T_1 = B_1^T B_1$, where B_1 is an upper bidiagonal matrix, and try to compute B_1 without touching T and T_1. This can be done as follows:

- Find the first column h of $T - sI = B^T B - sI$.

- Build up the rotation matrix G so that $Gh = [*, 0, \ldots, 0]^T$.

- Compute the matrix $W_0 = BG^T$, and, then, with the help of rotations of columns and rows turn it into an upper bidiagonal matrix $W_1 = ZW_0V$ (with Z, V being products of the rotations involved).

- Set $B_1 = W_1$.

As is readily seen, $B_1^T B_1 = (GV)^T (B^T B) (GV)$, and the QR decomposition of $T - sI$ can be done so that the first columns of the orthogonal matrices GV and Q coincide.

Exercises

1. Give an example of a matrix for which the QR algorithm with the Rayleigh shift (multishift of order 1) does not converge.

2. Suppose that B is an upper bidiagonal matrix, and consider the following process:

$$B_0 = B; \quad C_k = B_{k-1}Q_k, \quad B_k = Z_kC_k, \quad k = 1, 2, \ldots,$$

where B_k are upper bidiagonal, C_k are lower bidiagonal, and Q_k, Z_k are unitary matrices. Prove that the matrices B_k and C_k converge to a common diagonal matrix, as $k \to \infty$.

3. Devise an algorithm producing the upper and lower bidiagonal matrices of the previous problem. Show how it serves to compute the singular value decomposition of a bidiagonal matrix.

4. Devise an algorithm that computes the singular values of an arbitrary matrix A.

5. Let $A = A^T \in \mathbb{C}^{n \times n}$. Prove that there exists a decomposition of the form

$$A = V \Sigma V^T,$$

where V is a unitary matrix while Σ is a diagonal matrix with a non-negative diagonal. Propose an algorithm to compute it.

Lecture 12

12.1 Function approximation

Given a class of functions F and a subclass of "simple" functions $\Phi \subset F$, one often wishes to approximate $f \in F$ by some $\phi \approx f$, $\phi \in \Phi$.

To move from the intuitive description to a strict one, it is necessary to specify the classes F and Φ and give a rigorous definition to what is "to approximate". Let F be a linear space of functions defined on a domain Ω. Then, there are two approaches.

Minimization approach. Choose a norm (or a seminorm [1]) $\|\cdot\|$ on F, and seek for a function $\phi \in \Phi \subset F$ that minimizes $\|f - \phi\|$.

Interpolation approach. Choose some nodes $x_0, \ldots, x_n \in \Omega$, and seek for a function $\phi \in \Phi$ that satisfies the interpolative conditions

$$\phi(x_i) = f(x_i), \quad i = 0, 1, \ldots, n.$$

12.2 Polynomial interpolation

Consider functions $f(x)$ of one real variable x. If $f(x)$ is chosen to be approximated by a polynomial

$$\phi(x) = L_n(x) \equiv a_n x^n + a_{n-1} x^{n-1} + \ldots + a_0,$$

then the interpolative conditions take the form

$$\begin{bmatrix} 1 & x_0 & x_0^2 & \cdots & x_0^n \\ 1 & x_1 & x_1^2 & \cdots & x_1^n \\ \cdots & \cdots & \cdots & \cdots & \cdots \\ 1 & x_n & x_n^2 & \cdots & x_n^n \end{bmatrix} \begin{bmatrix} a_0 \\ a_1 \\ \cdots \\ a_n \end{bmatrix} = \begin{bmatrix} f(x_0) \\ f(x_1) \\ \cdots \\ f(x_n) \end{bmatrix}. \qquad (*)$$

---

[1] The seminorm differs from the norm in that the zero value does not imply that the element is zero.

This is a system of linear equations. Denote by W^T the coefficient matrix. Then $W = W(x_0, \ldots, x_n)$ is the Vandermonde matrix. It is well known that (prove it)

$$\det W(x_0, \ldots, x_n) = \prod_{0 \le i < j \le n} (x_j - x_i).$$

If the nodes $x_0, \ldots, x_n$ are pairwise distinct, then the coefficient matrix for $(*)$ is nonsingular. It follows that the interpolating polynomial exists and is determined uniquely.

It would seem that $L_n(x)$ could have been obtained by standard methods for solving $(*)$. However, nobody does so for two reasons:

- Standard methods ignore the structure of W^T.

- The Vandermonde matrix W is ill-conditioned:[2] for any pairwise distinct real nodes $x_0, x_1, \ldots, x_n$,

$$\text{cond}_2 \, W(x_0, x_1, \ldots, x_n) \ge 2^{n-2}/\sqrt{n}.$$

For particular configurations of nodes, there are some other estimates to boot:[3] if the nodes are positive, then $\text{cond}_1 \, W > 2^n$.

There is no obligation at all to compute *the coefficients* of the polynomial. We need but an easy way to compute the value of the polynomial at any prescribed point.

12.3 Interpolating polynomial of Lagrange

If polynomials $l_0(x), \ldots, l_n(x)$ of order n satisfy the interpolative conditions

$$l_j(x_i) = \left\{ \begin{array}{l} 1, i = j, \\ 0, i \ne j, \end{array} \right.$$

then, evidently,

$$L_n(x) \equiv \sum_{j=0}^{n} f(x_j) l_j(x). \tag{12.3.1}$$

The polynomials $l_j(x)$ exist and unique (for they are the solutions of a polynomial interpolative problem). These polynomials are called the elementary Lagrange polynomials.

A polynomial L_n is termed the *Lagrange interpolation polynomial,* and the polynomial interpolative problem itself is frequently referred to as the *Lagrange interpolation.*

---

[2]E.E.Tyrtyshnikov. How bad are Hankel matrices? *Numer. Math.* 67: 261–269 (1994).

[3]W.Gautschi and G.Inglese. Lower bounds for the condition number of Vandermonde matrices. *Numer. Math.* 52: 241–250 (1988).

It is easy to verify that

$$l_j(x) = \prod_{\substack{k=0 \\ k \neq j}}^{n} \frac{x - x_k}{x_j - x_k}. \qquad (12.3.2)$$

Taking this into account, we obtain a useful formula (check it!)

$$L_n(x) = \sum_{j=0}^{n} \frac{f(x_j)\, \omega(x)}{(x - x_j)\, \omega'(x_j)}, \qquad \omega(x) = \prod_{k=0}^{n} (x - x_k) \qquad (12.3.3)$$

($\omega'(x)$ is the derivative of $\omega(x)$ in x).

12.4 Error of Lagrange interpolation

Theorem 12.4.1 *Let* $x, x_0, \ldots, x_n \in [a,\, b]$ *and* $f \in C^{n+1}[a,\, b]$. *Then*

$$f(x) - L_n(x) = \frac{f^{(n+1)}(\xi(x))}{(n+1)!}\, \omega(x), \quad \omega(x) = \prod_{k=0}^{n} (x - x_k),$$

where

$$\min\{x, x_0, \ldots, x_n\} \;<\; \xi(x) \;<\; \max\{x, x_0, \ldots, x_n\}.$$

Proof. Fix $x \notin \{x_0, \ldots, x_n\}$. Then $\omega(x) \neq 0$, and we may consider the following function of t:

$$g(t) \equiv f(t) - L_n(t) - c\, \omega(t), \quad c \equiv \frac{f(x) - L_n(x)}{\omega(x)}.$$

The function $g(t)$ is zero at $t = x, x_0, \ldots, x_n$. By Rolle's theorem, $g^{(1)}(t)$ has $n+1$ zeroes $\Rightarrow g^{(2)}(t)$ has n zeroes $\Rightarrow \ldots \Rightarrow \exists\, \xi : g^{(n+1)}(\xi) = 0$. It remains to be noted that

$$g^{(n+1)}(t) = f^{(n+1)}(t) - c\,(n+1)!. \quad \square$$

12.5 Divided differences

The values of a function $f(x)$ at the nodes will be referred to as its divided differences of order 0. For any pair of nodes x_0, x_1, the quantities

$$f(x_0; x_1) \equiv \frac{f(x_1) - f(x_0)}{x_1 - x_0}$$

will be termed the divided differences of order 1. By induction, the quantities

$$f(x_0; \ldots; x_k) \equiv \frac{f(x_1; \ldots; x_k) - f(x_0; \ldots; x_{k-1})}{x_k - x_0}$$

will be called the divided differences of order k.

Lemma 12.5.1

$$f(x_0; \ldots; x_k) = \sum_{j=0}^{k} \frac{f(x_j)}{\prod\limits_{\substack{l=0 \\ l \neq j}} (x_j - x_l)}.$$

Proof. By induction,

$$f(x_0; \ldots; x_k) = \sum_{j=1}^{k} \frac{f(x_j)}{\prod\limits_{\substack{l=1 \\ l \neq j}}^{k} (x_j - x_l)(x_k - x_0)} - \sum_{j=0}^{k-1} \frac{f(x_j)}{\prod\limits_{\substack{l=0 \\ l \neq j}}^{k-1} (x_j - x_l)(x_k - x_0)}$$

$$= \frac{f(x_0)}{\prod\limits_{\substack{l=0 \\ l \neq 0}}^{k} (x_0 - x_l)} + \frac{f(x_k)}{\prod\limits_{\substack{l=0 \\ l \neq k}}^{k} (x_k - x_l)}$$

$$+ \sum_{j=1}^{k-1} \frac{f(x_j)}{\prod\limits_{\substack{l=1 \\ l \neq j}}^{k-1} (x_j - x_l)(x_k - x_0)} \left\{ \frac{1}{x_j - x_k} - \frac{1}{x_j - x_0} \right\}. \square$$

Corollary 12.5.1 *The value of a divided difference $f(x_0; \ldots; x_k)$ does not depend on the ordering of the nodes.*

Corollary 12.5.2 $f(x) - L_n(x) = f(x; x_0; \ldots; x_n)\, \omega(x).$

Proof.

$$f(x) - L_n(x) = \omega(x) \left\{ \frac{f(x)}{\omega(x)} + \sum_{j=0}^{n} \frac{f(x_j)}{(x_j - x)\, \omega'(x_j)} \right\}. \quad \square$$

Corollary 12.5.3

$$f(x_0; \ldots; x_k) = \frac{f^{(k)}(\xi)}{k!},$$

where

$$\min\{x_0, \ldots, x_k\} < \xi < \max\{x_0, \ldots, x_k\}.$$

12.6   Newton formula

The following formula holds and can be viewed as a discrete analog of the Taylor series:

$$\begin{aligned}
L_n(x) = f(x_0) \;&+\; f(x_0; x_1)(x - x_0) \\
&+\; f(x_0; x_1; x_2)(x - x_0)(x - x_1) \\
&+\; \ldots \\
&+\; f(x_0; x_1; \ldots; x_n)(x - x_0) \ldots (x - x_{n-1}).
\end{aligned}$$

Proof. Write

$$L_n = L_0 + (L_1 - L_0) + (L_2 - L_1) + \ldots + (L_n - L_{n-1}),$$

where L_k is the lagrange polynomial interpolating the function $f(x)$ at the nodes $x_0, \ldots, x_k$.

Since L_{k-1} interpolates L_k at the nodes $x_0, \ldots, x_{k-1}$,, we find that

$$L_k(x) - L_{k-1}(x) = L_k(x; x_0; \ldots; x_{k-1})(x - x_0) \ldots (x - x_{k-1}).$$

Setting $L_k(x) = a_k x^k + \ldots$, by Corollary 12.5.3, we infer that the quantity

$$L_k(x; x_0; \ldots; x_{k-1}) = a_k$$

does not depend on x. Hence, we may take $x = x_k$, and then

$$L_k(x_k; x_0; \ldots; x_{k-1}) = f(x_0; \ldots; x_k). \quad \square$$

In contrast to the classical Taylor series, its discrete counterpart possesses two advantages. First, it does not involve derivatives. Second, it can provide far better accuracy (we shall see this shortly).

12.7 Divided differences with multiple nodes

Suppose that there are equal (multiple) nodes among $x_0, \ldots, x_n$. If $y \in \{x_0, \ldots, x_n\}$ occurs precisely m times, then y is called the node of multiplicity m.

A suite of nodes $M = \{x_0, \ldots, x_n\}$ is termed a *multiple mesh*, if there are multiple nodes, and a *simple mesh* if all the nodes are pairwise distinct.

For any mesh $M = \{x_0, \ldots, x_n\}$, there exist a family of simple meshes $M^\varepsilon = \{x_0^\varepsilon, \ldots, x_n^\varepsilon\}$, $\varepsilon > 0$ providing that $x_i^\varepsilon \to x_i$ as $\varepsilon \to 0$ for all i. Then by a divided difference with multiple nodes is meant the limit

$$f(x_0; \ldots; x_n) \equiv \lim_{\varepsilon \to 0} f(x_0^\varepsilon; \ldots; x_n^\varepsilon). \qquad (*)$$

Lemma 12.7.1 *If the multiplicity for every node does not exceed m and $f \in C^{m-1}$, then the limit $(*)$ exists and does not depend on the choice of simple meshes M^ε.*

Proof. Divided differences for simple meshes do not depend on the ordering of nodes. Therefore, we may regard the nodes of the simple meshes to be ordered so that if any two nodes converge to a common point, then all the

nodes in between them converge to the same point. There is no question of
the existence of the divided differences $f(x_i)$ of order 0. Further,

$$
f(x_i^\varepsilon; x_{i+1}^\varepsilon) = \begin{cases} \frac{f(x_{i+1}^\varepsilon) - f(x_i^\varepsilon)}{x_{i+1}^\varepsilon - x_i^\varepsilon}, & x_i \neq x_{i+1}, \\[2ex] f^{(1)}(\xi^\varepsilon), & x_i = x_{i+1}. \end{cases}
$$

In the case $x_i \neq x_{i+1}$, for all sufficiently small ε, we have $x_i^\varepsilon \neq x_{i+1}^\varepsilon$, and,
thence, the limits (as $\varepsilon \to 0$) of the denominator and numerator exist. In the
case $x_i = x_{i+1}$, the limit exists owing to the property that ξ^ε lies between
x_i^ε and x_{i+1}^ε.

According to the recursive definition of divided differences for simple meshes,

$$
f(x_i^\varepsilon; \ldots; x_{i+k}^\varepsilon) = \begin{cases} \frac{f(x_{i+1}^\varepsilon; \ldots; x_{i+k}^\varepsilon) - f(x_i^\varepsilon; \ldots; x_{i+k-1}^\varepsilon)}{x_{i+k}^\varepsilon - x_i^\varepsilon}, & x_i \neq x_{i+k}, \\[2ex] f^{(k)}(\xi^\varepsilon), & x_i = x_{i+k}. \end{cases}
$$

Suppose that the limits for the divided differences of order $k-1$ exist. Then
for $x_i \neq x_{i+k}$, there exist the limits for the numerator and denominator. For
$x_i = x_{i+k}$ the limit also exists, because

$$
\min\{x_i^\varepsilon, \ldots, x_{i+k}^\varepsilon\} < \xi^\varepsilon < \max\{x_i^\varepsilon, \ldots, x_{i+k}^\varepsilon\},
$$

which implies that $\xi^\varepsilon \to x_i = x_{i+1}$ as $\varepsilon \to 0$. $\square$

Corollary 12.7.1 *If $f \in C^k$, then*

$$
f(x_0; \ldots; x_k) = \frac{f^{(k)}(\xi)}{k!}, \tag{12.7.1}
$$

where

$$
\min\{x_0, \ldots, x_k\} \leq \xi \leq \max\{x_0, \ldots, x_k\}. \tag{12.7.2}
$$

Proof. In the case of simple nodes, the proof stems from Corollary 12.5.3.
In the case of multiple nodes, $f(x_0; \ldots; x_k)$ is the limit of the quantities

$$
f(x_0^\varepsilon; \ldots; x_k^\varepsilon) = \frac{f^{(k)}(\xi^\varepsilon)}{k!}.
$$

Let $\varepsilon = 1/N$. Loosely speaking, the right-hand side is not bound to converge
as $N \to \infty$. For all this, some subsequence of points ξ^ε is granted to converge
to some point ξ satisfying the inequalities (12.7.2). $\square$

12.8   Generalized interpolative conditions

If a node $z \in \{x_0, \ldots, x_n\}$ is of multiplicity m, then the generalized interpolating polynomial $H_n(x)$ of order n (or less) is defined by the following
generalized interpolative conditions:

$$
H_n^{(j)}(z) = f^{(j)}(z), \quad j = 0, \ldots, m-1.
$$

If there are ν pairwise distinct nodes $z_1, \ldots, z_\nu$ of multiplicities $m_1, \ldots, m_\nu$, then

$$n = m_1 + \ldots + m_\nu - 1.$$

It is not difficult to establish the uniqueness of the polynomial H_n. (Do this!) The following generalization of the Newton formula takes place:

$$
\begin{aligned}
H_n(x) = f(x_0) \quad &+ \quad f(x_0; x_1)(x - x_0) \\
&+ \quad f(x_0; x_1; x_2)(x - x_0)(x - x_1) \\
&+ \quad \ldots \\
&+ \quad f(x_0; x_1; \ldots; x_n)(x - x_0) \ldots (x - x_{n-1}).
\end{aligned}
$$

Proof. Consider simple meshes M^ε and the corresponding Lagrange polynomials $L_n^\varepsilon(x)$. Since $L_n^\varepsilon(x) \to H_n(x)$ as $\varepsilon \to 0$, we obtain the same polynomial $H_n(x)$ for any ordering of nodes. Let $z = x_0 = \ldots = x_{m-1}$ be a node of multiplicity m. Then

$$H_n(x) = \sum_{j=0}^{m-1} \frac{f^{(j)}(z)}{j!}(x - z)^j + (x - z)^m \, p_{n-m}(x),$$

where the degree of the polynomial $p_{n-m}(x)$ is less than or equal to $n - m$. The generalized interpolative conditions at the node z are met obviously. $\quad\square$

Prove that if $f \in C^{n+1}$ then

$$f(x) - H_n(x) = \frac{f^{n+1}(\xi(x))}{(n+1)!} \, w(x), \quad w(x) = \prod_{k=0}^{n}(x - x_k),$$

where

$$\min\{x, x_0, \ldots, x_n\} \le \xi(x) \le \max\{x, x_0, \ldots, x_n\}.$$

The generalized interpolation polynomial is often referred to as *the Hermitian polynomial* while the generalized interpolation problem is talked about as *the Hermitian interpolation*.

12.9 Table of divided differences

When calculating the divided differences in the case of simple and multiple nodes, explicitly or implicitly, one builds the following table of divided differences:

$$f(x_0)$$

$$f(x_1) \quad \cdots \quad f(x_0; x_1)$$

$$f(x_2) \quad \cdots \quad f(x_1; x_2) \quad \cdots \quad f(x_0; x_1; x_2)$$

$$f(x_3) \quad \cdots \quad f(x_2; x_3) \quad \cdots \quad f(x_1; x_2; x_3) \quad \cdots \quad f(x_0; x_1; x_2; x_3)$$

$$\cdots \quad \cdots \quad \cdots \quad \cdots \quad \cdots \quad \cdots \quad \cdots \; .$$

In the case of multiple nodes, one should list the nodes so that the equality $x_i = x_j$ entails $x_k = x_i$ for all $i \le k \le j$.

The diagonal of the table of divided differences contains the coefficients of the discrete analog of the Taylor series. Once we have obtained them, we can easily solve the Lagrange and Hermitian interpolative problems.

Exercises

1. Consider the following table of values for a polynomial of second degree:

x	-2	-1	0	1	2
$f(x)$	7	3	1	0	3

There is exactly one error in the second row. Find the error, correct it, and recover the polynomial.

2. One makes a table of values for the function

$$f(x) = \frac{2}{\pi} \int\limits_0^x e^{-t^2} dt$$

on the interval $[0, 1]$ with a constant step h. It is required that the quadratic interpolation error must not exceed 0.01. What must h be?

3. A polynomial $f(x) = x^n + a_1 x^{n-1} + \ldots + a_n$ has pairwise distinct roots $x_1, \ldots, x_n$. Prove that

$$\sum_{j=1}^{n} \frac{x_j^k}{f'(x_j)} = \begin{cases} 0, & 0 \le k \le n-2, \\ 1, & k = n-1. \end{cases}$$

4. Prove that if $f \in C^k$, then

$$f(x_0; \ldots; x_k) = \int_0^1 \int_0^{t_1} \cdots \int_0^{t_{k-1}} f^{(k)}(x_0 + t_1(x_1 - x_0) + \cdots + t_k(x_k - x_{k-1})) \, dt_1 \, dt_2 \ldots dt_k \quad .$$

5. Prove the uniqueness of the polynomial solving the Hermitian interpolation problem.

6. The Lagrange polynomial $L_n(x)$ approximates a function $f(x)$ with error $\varepsilon > 0$. With what error does $L'_n(x)$ approximate $f'(x)$?

7. Given the values of a function $f(x)$ at pairwise distinct points $x_1, \ldots x_n$, the coefficients of the interpolating polynomial are required. Propose an algorithm that does this in $\mathcal{O}(\log_2 n)$ parallel steps.

8. Suppose that for some $\alpha_1, \ldots \alpha_n$ we are required to find a function of the form $\phi(x) = c_1 e^{\alpha_1 x} + \ldots + c_k e^{\alpha_n x}$ which interpolates the values of $f(x)$ at pairwise distinct nodes $x_1, \ldots x_n$. What are the conditions on $\alpha_1, \ldots \alpha_n$ under which this problem has a unique solution?

9. An analytic function $f(x) = \sum\limits_{k=0}^{\infty} a_k x^k$ is given, and it is required to find a function of the form

$$\phi(x) = \frac{p_0 + p_1 x + \ldots + p_{n-1} x^{n-1}}{q_0 + q_1 x + \ldots + q_n x^n}$$

satisfying the generalized interpolative conditions:

$$\phi^{(j)}(0) = f^{(j)}(0), \quad j = 0, 1, \ldots, 2n.$$

Prove that if the matrices $A_k = [a_{k+i-j}]_{i,j=0}^{k}$ are nonsingular for $k = n$ and $k = n - 1$, then the solution exists.

Lecture 13

13.1 Convergence of the interpolation process

Consider a sequence of simple meshes on an interval $[a, b]$

$$M_n = \{x_{n0}, \ldots, x_{nn}\}, \quad n = 0, 1, \ldots$$

and the sequence of the Lagrange polynomials $L_n(x)$ interpolating the values of $f(x)$ at the nodes of M_n.

Is it true that $L_n(x) \to f(x) \;\; \forall \;\; x \in [a, b]$? Is the convergence uniform in x? How do we estimate the convergence rate?

The answers depend on some properties of the sequence M_n and the function $f(x)$. Let $f \in C^\infty [a, b]$, and, for some number $M > 0$, assume that $\sup_x |f^{(n)}(x)| \leq M^n$ for all n. Then Theorem 12.4.1 on the Lagrange interpolation error provides that

$$\|f - L_n\|_{C [a, b]} \equiv \max_{a \leq x \leq b} |f(x) - L_n(x)| \leq \frac{(M (b - a))^{n+1}}{(n + 1)!}.$$

The right-hand side, obviously, tends to zero as $n \to \infty$.

In a sense, the less we demand of f, the more we require of the meshes.

13.2 Convergence of the projectors

Let $F = C[a, b]$ be a Banach space with the norm

$$\|f\|_{C [a, b]} \equiv \max_{a \leq x \leq b} |f(x)|.$$

If Π_n is the space of polynomials of degree n or less, then the Lagrange interpolation for $f \in F$ defines the operator

$$P_n : F \to \Pi_n, \quad P_n f = L_n.$$

P_n is a linear continuous operator. (Why?) Moreover, it is a projector, because $P_n^2 = P_n$.

The question of the interpolation process convergence can be cast as follows: concerning the property

$$P_n f \to f \quad \forall f \in F,$$

under what conditions does it hold?

13.3 Sequences of linear continuous operators in a Banach space

Consider a Banach space F and a sequence of linear continuous operators $P_n : F \to F$ (not necessarily projectors).

Theorem 13.3.1 (Uniform boundedness principle)

$$\sup_n \|P_n f\| \leq c(f) < +\infty \quad \forall f \in F \quad \Longleftrightarrow \quad \sup_n \|P_n\| < +\infty.$$

Proof. The part "$\Leftarrow$" is evident. We prove the part "$\Rightarrow$" by contradiction. Use the following (rather lucid) observation: if a sequence $\|P_n\|$ is not bounded, then in any ball there exist some elements u_n on which $\|P_n u_n\| \to \infty$.
    For $m = 1, 2, \ldots$, consider the sets

$$O_m \equiv \{f : \sup_n > m\}.$$

These are open sets (why?) such that one is embedded into the other: $O_1 \supset O_2 \supset \ldots$.
    Assume that O_m is not empty for all m. Then it is dense in any ball. Consider open and close balls B_m and $\bar{B}_m$ of radius $r_m \leq 1/m$, belonging to $O_m \cap B_{m-1}$ (set $B_0 \equiv O_1$). The balls $\bar{B}_m$ make up a sequence of nested closed balls with the radii tending to zero. Hence,

$$\exists\, f \in \bigcap_{m=1}^{\infty} \bar{B}_m \subset \bigcap_{m=1}^{\infty} O_m \quad \Longrightarrow \quad \sup_n \|P_n f\| > m \quad \forall\, m. \quad \square$$

Theorem 13.3.2 (Banach–Steinhaus.) *For a sequence of functions $P_n f$ to be convergent for all $f \in F$, it is necessary and sufficient that*

(1) $\sup_n \|P_n\| < +\infty$, *and*

(2) $P_n f$ *is convergent, as $n \to \infty$, on a subset $\tilde{F}$ which is dense in F.*

Proof. The "necessary" assertion emanates from the uniform bounded-ness principle. To prove the "sufficient" one, consider $f \in F$ and an ε-approximation $f_\varepsilon \in \tilde{F}$: $\|f - f_\varepsilon\| \le \varepsilon$. Then

$$\begin{aligned} \|P_n f - P_m f\| &\le \|P_n f - P_n f_\varepsilon\| + \|P_n f_\varepsilon - P_m f_\varepsilon\| + \|P_m f_\varepsilon - P_m f\| \\ &= \mathcal{O}(\varepsilon) \quad \text{for} \quad n, m \to \infty. \quad \square \end{aligned}$$

If P_n are the interpolative projectors, then the convergence on a set dense everywhere in $C[a, b]$ is taken for granted. (Why?) For the convergence analysis of the interpolation process, the key question now becomes the following: how do the norms $\|P_n\|$ behave, as $n \to \infty$?

13.4 Algebraic and trigonometric polynomials

A change of variable $x = \frac{a+b}{2} + \frac{b-a}{2}t$ allows one to get on to functions of $t \in [-1, 1]$.

One more change of variable $t = \cos \phi$ allows one to turn to 2π-periodic functions which are defined on the whole real axis and even. An algebraic polynomial $p_n(t)$ of order n or less converts into an even trigonometric polynomial

$$p_n(\cos \phi) = q_n(\phi) \equiv \sum_{k=0}^{n} \alpha_k \cos k\phi.$$

It can be proved (by induction) that a function $\cos k\phi$ is an algebraic polynomial in $\cos \phi$ of degree k. Therefore, any even trigonometric polynomial $q_n(\phi)$ of order n or less can be written in the form $p_n(\cos \phi)$, where $p_n(t)$ is an algebraic polynomial of degree n or less.

An arbitrary (not necessarily even) trigonometric polynomial of order n or less is defined as

$$Q_n(\phi) = \alpha_0 + \sum_{k=1}^{n} (\alpha_k \cos k\phi + \beta_k \sin k\phi),$$

and it is also convenient to write it in the form

$$Q_n(\phi) = \sum_{k=-n}^{n} c_k e^{ik\phi},$$

where $c_{-k} = \bar{c}_k$ (the complex conjugate) for all k (in this and only in this case the values of $Q_n(\phi)$ are real for all ϕ).

An important conclusion: all questions about the function approximation by algebraic polynomials reduce to those about the approximation of even periodic functions by even trigonometric polynomials.

13.5   The Fourier series projectors

Let C be a Banach space of real continuous 2π-periodic functions with the norm of the space $C[-\pi, \pi]$, and let S_n be a projector giving $f \in C$ to go into the n-term partial sum of its Fourier series:

$$s_n(\phi) = \sum_{k=-n}^{n} c_k\, e^{i\,k\,\phi},$$

where

$$c_k = \frac{1}{2\pi} \int_{-\pi}^{\pi} f(\psi)\, e^{-i\,k\,\psi}\, d\psi.$$

Lemma 13.5.1 $\|S_n\| \geq c \ln n, \quad c > 0.$

Proof. Allowing for the expressions for c_k, we find that

$$s_n(\phi) = \frac{1}{2\pi} \int_{-\pi}^{\pi} \left(\sum_{k=-n}^{n} e^{i\,k\,(\phi-\psi)} \right) f(\psi)d\psi = \frac{1}{2\pi} \int_{-\pi}^{\pi} \frac{\sin\left(n+\frac{1}{2}\right)t}{\sin\frac{t}{2}} f(\phi+t)dt.$$

Consequently,

$$\|S_n\| = \frac{1}{2\pi} \int_{-\pi}^{\pi} \left| \frac{\sin\left(n+\frac{1}{2}\right)t}{\sin\frac{t}{2}} \right| dt = \frac{1}{\pi} \int_{0}^{\pi} \left| \frac{\sin N u}{\sin u} \right| du \quad (N = 2n+1)$$

$$> \frac{1}{\pi} \sum_{k=0}^{N-1} \int_{\frac{\pi k}{N}}^{\frac{\pi k}{N} + \frac{\pi}{2N}} \left| \frac{\sin N u}{\sin u} \right| du = \frac{1}{\pi} \sum_{k=0}^{N-1} \int_{0}^{\frac{\pi}{2N}} \left| \frac{\sin N v}{\sin\left(\frac{\pi k}{N} + v\right)} \right| dv$$

$$> \frac{1}{\pi} \sum_{k=0}^{N-1} \int_{0}^{\frac{\pi}{2N}} \frac{\frac{2}{\pi} N v}{\frac{\pi k}{N} + v} dv > \frac{1}{\pi} \sum_{k=0}^{N-1} \frac{\frac{2}{\pi} N \left(\frac{\pi}{2N}\right)^2 \frac{1}{2}}{\frac{\pi k}{N} + \frac{\pi}{2N}}$$

$$= \frac{1}{4\pi} \sum_{k=0}^{N-1} \frac{1}{k + \frac{1}{2}} > \frac{1}{4\pi} \sum_{k=1}^{N-1} \frac{1}{k} \geq c \ln n. \quad \square$$

Corollary 13.5.1 *There exists a continuous periodic function for which the Fourier series does not converge uniformly to this or any other continuous function.*

Corollary 13.5.2 *For any point there exists a continuous periodic function for which the Fourier series diverges at this point.*

For any given point $x_0 \in [-\pi, \pi]$, it is sufficient to regard the functionals $s_n(x_0)$ and observe that $\|s_n(x_0)\| = \|S_n\|$.

13.6 "Pessimistic" results

Theorem 13.6.1 (Losinsky–Kharshiladze.) *Suppose a linear continuous operator $T_n : C \to C$ is a projector onto a subspace of trigonometric polynomials of degree n or less. Then*

$$||T_n|| \geq ||S_n|| \geq c \ln n.$$

Proof. We fall back on the *Zigmund–Martsinkevich–Berman identity*:

$$\frac{1}{2\pi} \int_{-\pi}^{\pi} [T_n \, f \, (t + h)] \, (t - h) \, d h \;=\; [S_n \, f] \, (t).$$

For $f(t) = \sin kt$ or $f(t) = \cos kt$, the identity is verified straightforwardly. Therefore, it holds true for arbitrary trigonometric polynomials, and then we use the fact that the set of trigonometric polynomials is dense everywhere in C.

Consider a function $f \in C$ such that

$$||S_n|| \;=\; ||S_n \, f||, \quad ||f|| \;=\; 1.$$

Obviously, $||f \, (t + h)|| \;=\; 1$ and $||T_n \, f \, (t + h)|| \; \leq \; ||T_n||$. Hence,

$$|[S_n \, f] \, (t)| \; \leq \; ||T_n||. \qquad \square$$

Corollary 13.6.1 (Faber's theorem) *For any sequence of simple meshes on $[a, b]$, there exists a function continuous on $[a, b]$ for which the sequence of interpolating polynomials does not converge uniformly to this or whatever else continuous function.*

A sequence of interpolating polynomials cannot converge uniformly in $f \in C\,[a, b]$, because $||P_n|| \to \infty$.

Try to show that there also exists a continuous function for which the interpolating polynomials diverge at least at one point.

Bernstein's example: if $f(x) \;=\; |x|$, $\quad x \in [-1, 1]$, then the interpolating polynomials on the uniform meshes will not converge to $f(x)$ at any point save for $x \;=\; -1, 0, 1$.

13.7 Why the uniform meshes are bad

For uniform meshes, the norms $||P_n||$ grow exponentially. Indeed, consider a uniform mesh on $[-1, 1]$:

$$t_j \;=\; -1 + \frac{2}{n} \, j, \quad j = 0, 1, \ldots, n.$$

Then

$$\|P_n\| = \max_{-1 \le t \le 1} \sum_{k=0}^{n} \left| \prod_{\substack{j=0 \\ j \ne k}}^{n} \frac{t - t_j}{t_k - t_j} \right|.$$

For $t = -1 + \frac{2}{n}\theta$, $0 < \theta < 1$, we find that

$$\|P_n\| \ge \sum_{k=0}^{n} \left| \prod_{\substack{j=0 \\ j \ne k}}^{n} \frac{\theta - j}{k - j} \right| \ge \frac{\theta(1 - \theta)}{n^2} \sum_{k=0}^{n} \frac{n!}{k!\,(n-k)!} = \frac{\theta(1 - \theta)}{n^2} 2^n,$$

since $2^n = \binom{n}{0} + \binom{n}{1} + \dots + \binom{n}{n}$.

We could obtain an even neater estimate:

$$\|P_n\| \ge \frac{\theta}{n} \sum_{k=0}^{n} \frac{n!}{k!\,(n-k)!} \prod_{k=1}^{n} (1 - \frac{\theta}{k}) \ge c\,\frac{2^n}{n^{1+\theta}}, \quad 0 < \theta < 1,$$

which follows from the gamma-function theory relationship

$$\prod_{k=1}^{n} (1 - \frac{\theta}{k}) = \frac{n^{-\theta}}{\Gamma(1 - \theta)} + \mathcal{O}(n^{-\theta-1}),$$

where

$$\Gamma(z) = \int_{0}^{\infty} x^{z-1} e^{-x}\,dx$$

(the Euler formula for the gamma-function).

13.8   Chebyshev meshes

To obtain "optimistic" results, we should give less room to constructing meshes. We cannot avoid the logarithmic growth of the norms $\|P_n\|$. But to prevent any higher growth, we need to use the *Chebyshev meshes*.

Take up a uniform mesh on $[-\pi, \pi]$ with $2n$ nodes. With trigonometric polynomials of degree not higher than $n - 1$, we can interpolate the values of any 2π-periodic function on this mesh. If the function is even, then the corresponding trigonometric polynomial will be even. Hence, it can be written in the form (why?)

$$T_{n-1}(\phi) = L_{n-1}(\cos \phi),$$

where $L_{n-1}(t)$ is an algebraic polynomial in t of degree not higher than $n - 1$.

A uniform mesh on $[-\pi, \pi]$ with $2n$ nodes induces some uniform mesh on $[0, \pi]$ with n nodes. Through the mapping $t = \cos \phi$, this mesh generates the *Chebyshev mesh* on $[-1, 1]$:

$$t_{nj} = \cos \left(\frac{\pi}{2n} + \frac{\pi}{n} j \right), \quad j = 0, 1, \dots, n - 1.$$

13.9   Chebyshev polynomials

The above quantities t_{nj} turn out to be the roots of the famous *Chebyshev polynomials* $T_n(t)$ defined by the recurrence

$$T_{n+1}(t) = 2T_n(t) - T_{n-1}(t), \qquad T_0 = T_1 = 1.$$

Lemma 13.9.1 *For any $0 \le t \le 1$,*

$$T_n(t) = \cos(n \ \arccos t) = 2^{n-1} \prod_{j=0}^{n-1} (t - t_{nj}).$$

Proof. Prove by induction that the function $C_n(t) \equiv \cos(n \ \arccos t)$ is, indeed, a polynomial in t. For $n = 0$ and $n = 1$, it is trivial. For $n \ge 1$,

$$\cos((n+1) \ \arccos t) + \cos((n-1) \ \arccos t) = 2\cos(n \ \arccos t)\cos(\arccos t)$$

$$\Rightarrow \qquad C_{n+1}(t) = 2\,C_n(t) - C_{n-1}(t) \qquad \Rightarrow \qquad C_n(t) = T_n(t).$$

It easy to see that T_n is a polynomial of degree n with the senior coefficient equal to 2^{n-1}. It remains to verify that the roots of T_n are precisely the quantities t_{nj}.   □

Corollary 13.9.1 *Let $P_n : \ C[-1, 1] \ \to \ C[-1, 1]$ be the interpolation projector for the Chebyshev mesh with $n + 1$ nodes. Then*

$$\|P_n\| = \max_{-1 \le t \le 1} \rho(t),$$

where

$$\rho(t) \equiv \sum_{k=0}^{n} \frac{|\cos((n+1) \ \arccos t)| \ \sin \frac{\pi(2k+1)}{2(n+1)}}{(n+1)\,|t - t_k|}.$$

13.10   Bernstein's theorem

Theorem 13.10.1 *For the interpolation projectors on the Chebyshev meshes,*

$$\|P_n\| = \mathcal{O}(\ln n).$$

Proof. It is sufficient to consider $t \in [0, 1]$. Let

$$t = \cos \frac{\pi(2m + 12\,\theta)}{2(n+1)}, \qquad 0 < \theta < \frac{1}{2}, \qquad m \text{ is integer.}$$

Then

$$\rho(t) = \sum_{k=0}^{n} \frac{|\cos \frac{\pi(2m+1+2\,\theta)}{2}| \ \sin \frac{\pi(2k+1)}{2(n+1)}}{(n+1)\,2\,|\sin \frac{\pi(k+m+1+\theta)}{2(n+1)}| \ |\sin \frac{\pi(k-m-\theta)}{2(n+1)}}$$

$$\le c \sum_{k=0}^{n} \frac{\theta(2k+1)}{|k+m+1+\theta| \ |k-m-\theta|}$$

(we use $c_1 \, \psi \leq \sin \psi \leq \psi$ for $0 \leq \psi \leq c_2 \, \pi$, $0 < c_2 < 1$)

$$\leq c \;\; + \;\; c \sum_{k=0}^{m-1} \frac{\theta \, (2k+1)}{(k+m+1+\theta)\,(m-k+\theta)}$$

$$+ \;\; \sum_{k=m+1}^{n} \frac{\theta \, (2k+1)}{(k+m+1+\theta)\,(k-m-\theta)}$$

$$= c \;\; + \;\; c\theta \sum_{k=0}^{m-1} \left(\frac{1}{m-k+\theta} - \frac{1}{k+m+1+\theta} \right)$$

$$+ \;\; c\theta \sum_{k=m+1}^{n} \left(\frac{1}{k+m+1+\theta} + \frac{1}{k-m-\theta} \right)$$

$$= \mathcal{O} \, (\ln n). \qquad \square$$

13.11 "Optimistic" results

Theorem 13.11.1 *If $f \in C^m\,[-1, 1]$, $m \geq 1$, and P_n is the interpolation projector on the Chebyshev mesh with $n+1$ nodes, then*

$$\|f - P_n \, f\|_{C\,[-1,\,1]} \;=\; \mathcal{O} \left(\frac{\ln n}{n^m} \right). \tag{13.11.1}$$

Consider a one-to-one correspondence

$$\Theta : \, f\,(t) \;\rightarrow\; g(\phi) \;\equiv\; f\,(\cos \phi).$$

Important that if $f \in C^m\,[-1, 1]$, then $g \in C^m\,[-\infty, \infty]$.
  There are many ways to approximate $g \in C^m$ by an even trigonometric polynomial $S_n \, g$ of degree not higher than n. It can be chosen so that

$$\|g - S_n \, g\|_C \;=\; \mathcal{O}\,(\frac{1}{n^m}). \tag{13.11.2}$$

This assertion is not trivial. However, if $g \in C^{m+1}$, then we can take as $S_n \, g$ the n-degree truncation of the Fourier series for g. Then (13.11.2) gets quite elementary.
  On the strength of (13.11.2),

$$\|f - \hat{S}_n \, f\|_{C\,[-1,\,1]} \;=\; \mathcal{O}\,(\frac{1}{n^m}), \quad \hat{S}_n \;=\; \Theta^{-1} \, S_n \, \Theta.$$

The function $\hat{S} f$ is an algebraic polynomial of degree n or less $\;\Longrightarrow\; P_n \hat{S}_n f = \hat{S}_n f$. Consequently,

$$\|f - P_n \, f\| \;\leq\; \|f - S_n \, f\| \,+\, \|P_n \, S_n \, f - P_n \, f\| \;\leq\; (1 + \|P_n\|) \, \|f - S_n \, f\|,$$

and then we appeal to Bernstein's estimate on the norms $\|P_n\|$.

Exercises

1. Prove the existence and uniqueness of the trigonometric polynomial of the form

$$Q_n(\phi) = \alpha_0 + \sum_{k=1}^{n} (\alpha_k \cos k\phi + \beta_k \sin k\phi)$$

   that interpolates the values on a simple mesh with $2n+1$ nodes on the interval $[-\pi, \pi]$.

2. Let $f(x) = |x|$ and $L_n(x)$ be the Lagrange polynomials on the Chebyshev meshes on $[-1, 1]$. Prove that $L_n(x) \to f(x)$ uniformly in $x \in [-1, 1]$.

3. Suppose that an analytic function

$$f(x) = \sum_{k=0}^{\infty} a_k x^k$$

   is interpolated by the Lagrange polynomials $L_n(x)$ on the Chebyshev meshes on $[-1, 1]$. Prove that for some $0 < q < 1$,

$$\|f - L_n\|_{C[-1,\,1]} = \mathcal{O}(q^n).$$

4. Let C be a Banach space of real continuous 2π-periodic functions with the norm of the space $C[-\pi, \pi]$ and let S_n be the projector causing $f \in C$ to go into the n-degree truncation of its Fourier series. Prove that

$$\|S_n\| \leq c \ln n,$$

   where $c > 0$ does not depend on n.

Lecture 14

14.1 Splines

A natural route to "optimistic" results for convergence of the interpolation process is to give up using "pure" algebraic polynomials and interpolate, for example, by piecewise polynomial functions. Such functions are referred to as *splines*. A spline is said to be of degree m if the degree of each polynomial is not higher than m and equal to m for at least one polynomial.

For a given simple mesh

$$a = x_0 < \ldots < x_n = b$$

and values $f_k = f(x_k)$, consider the set of functions

$$\Phi \equiv \{\phi \in C^2[a, b] : \phi(x_k) = f_k, \; k = 0, 1, \ldots, n\}. \tag{14.1.1}$$

A function $S(x) \in \Phi$ is termed the *interpolative cubic spline* if on every interval $[x_{k-1}, x_k]$, $k = 1, \ldots, n$, it is a cubic polynomial.

14.2 Natural splines

To build up a cubic spline, we need to find $4n$ coefficients defining cubic polynomials on every interval. From the definition of the interpolative cubic spline we obtain $4n - 2$ equations (check this). For the number of equations to coincide with the number of unknowns, we may impose two extra conditions.

An interpolative cubic spline subject to the extra conditions

$$S''(x_0) = S''(x_n) = 0 \tag{14.2.1}$$

is said to be a *natural spline* (for Φ).

One has to deal also with "unnatural" interpolation cubic splines. For instance, instead of (14.2.1), we can interpolate the first-order derivatives

$$S'(x_0) = f'(x_0), \quad S'(x_n) = f'(x_n), \tag{14.2.2}$$

or (in the case $f_0 = f_n$) impose periodicity conditions on the first- and second-order derivatives:

$$S'(x_0) = S'(x_0), \quad S''(x_n) = S''(x_n). \tag{14.2.3}$$

14.3 Variational property of natural splines

An imposing property of the natural spline: it minimizes the "energy" functional

$$E(\phi) \equiv \int_a^b (\phi''(x))^2 \, dx.$$

Theorem 14.3.1 *Suppose that Φ is of the form (14.1.1) and $S(x)$ is a natural spline for Φ. Then*

$$E(\phi) \geq E(S) \quad \forall \, \phi \in \Phi,$$

and the inequality is strict whenever $\phi \neq S$.

Proof. $(\phi'')^2 - (S'')^2 = (\phi'' - S'')^2 + 2 S'' (\phi'' - S'') \quad \Rightarrow$

$$E(\phi) - E(S) = E(\phi - S) + 2 \int_a^b S'' (\phi'' - S'') \, dx.$$

Using integration by parts, we obtain

$$\int_a^b S'' (\phi'' - S'') \, dx. = \sum_{k=1}^n (S'' (\phi' - S') |_{x_{k-1}}^{x_k}$$
$$- S''' (x_{k-1} + 0) (\phi - S) |_{x_{k-1}}^{x_k}) = 0. \quad \square$$

14.4 How to build natural splines

It is time to make sure of the existence and uniqueness of natural splines.

Consider $S(x)$ on the kth interval $[x_{k-1}, x_k]$ and set

$$u_k = S''(x_k), \quad h_k = x_k - x_{k-1}, \quad x = x_{k-1} + t h_k.$$

Since $S(x)$ is a cubic polynomial as $x \in [x_{k-1}, x_k]$, we find the following:

$$S''(x) = (1-t) u_{k-1} + t u_k \quad \Rightarrow$$
$$S'(x) = S'(x_{k-1}) + h_k \left(\frac{1}{2} - \frac{(1-t)^2}{2} \right) u_{k-1} + h_k \frac{t^2}{2} u_k \quad \Rightarrow$$
$$S(x) = S(x_{k-1}) + h_k t S'(x_{k-1})$$
$$+ h_k^2 \left(\frac{t}{2} + \frac{(1-t)^3}{6} - \frac{1}{6} \right) u_{k-1} + h_k^2 \frac{t^3}{6} u_k.$$

Set $\delta f_k = (f_k - f_{k-1})/h_k$. Then for $t = 1$ the last equation gives

$$S'(x_{k-1}) = \delta f_k - \frac{h_k}{3} u_{k-1} - \frac{h_k}{6} u_k.$$

Allowing for this, we set $t = 1$ in the expression for $S'(x)$:

$$S'(x_k) = \delta f_k + \frac{h_k}{6} u_{k-1} + \frac{h_k}{3} u_k.$$

Equating the first-order derivatives of the spline on the right end of the interval k and on the left end of the interval $k + 1$, we obtain the equation

$$\delta f_k + \frac{h_k}{6} u_{k-1} + \frac{h_k}{3} u_k = \delta f_{k+1} - \frac{h_{k+1}}{3} u_k - \frac{h_{k+1}}{6} u_{k+1} \quad \Rightarrow$$

$$h_k u_{k-1} + 2 (h_k + h_{k+1}) u_k + h_{k+1} u_{k+1} = \rho_k \equiv 6 (\delta f_{k+1} - \delta f_k).$$

Since $u_0 = u_n = 0$ we have obtained, indeed, a linear system

$$T \begin{bmatrix} u_1 \\ \dots \\ u_{n-1} \end{bmatrix} = \begin{bmatrix} \rho_1 \\ \dots \\ \rho_{n-1} \end{bmatrix}$$

with the tridiagonal coefficient matrix

$$T = \begin{bmatrix} 2(h_1 + h_2) & h_2 & & \\ h_2 & 2(h_2 + h_3) & h_3 & \\ & \dots & \dots & \dots \\ & & h_{n-1} & 2(h_{n-1} + h_n) \end{bmatrix}. \tag{14.4.1}$$

Theorem 14.4.1 *The natural spline exists and is determined uniquely.*

Proof. We substitute $S'(x_{k-1})$ into the expression for $S(x)$ and so arrive at

$$S(x) = (1 - t) f_{k-1} + t f_k + u_{k-1} h_k^2 a(t) + u_k h_k^2 b(t), \tag{14.4.2}$$

where

$$a(t) = \frac{t}{6} + \frac{(1-t)^3}{6} - \frac{1}{6}, \quad b(t) = -\frac{t}{6} + \frac{t^3}{6}. \tag{14.4.3}$$

For any u_k, formula (14.4.2) gives a piecewise polynomial function interpolating the values f_k and possessing a continuous second-order derivative. The continuity property for the first-order derivative is described by the system (14.4.1 with respect to u_k. The coefficient matrix T is diagonally dominant and, thence, nonsingular. $\square$

14.5   Approximation properties of natural splines

Consider simple meshes M_n $a = x_{n0} < \ldots < x_{nn} = b$ and on them the natural splines interpolating the values of a function $f(x)$. Let

$$\Delta_n \equiv \max_k h_k, \quad \delta_n \equiv \min_k h_k.$$

A sequence of meshes M_n is called *quasi-uniform* if $\Delta_n/\delta_n \leq c < +\infty$.

Theorem 14.5.1 *For any function $f(x) \in C[a, b]$, a sequence of natural splines $S_n(x)$ on quasi-uniform meshes M_n converges to $f(x)$ uniformly in $x \in [a, b]$, provided that $\Delta_n \to 0$.*

Proof. Let u_i be the maximal in modulus component of the vector u. From the ith equation of the system $Tu = \rho$, we find that

$$|u_i| \leq |\rho_i|/(h_i + h_{i+1}) \quad \Rightarrow$$

$$\|u\|_\infty \leq c_1 \frac{\omega(\Delta_n; f)}{\delta_n^2}, \quad c_1 > 0,$$

where

$$\omega(\Delta; f) \equiv \max_{|x-y| \leq \Delta} |f(x) - f(y)|$$

is the continuity modulus of the function f. According to (14.4.2),

$$|S(x) - f(x)| \leq \omega(\Delta_n; f)\left(1 + c_1 \frac{\Delta_n^2}{\delta_n^2}(|a(t)| + |b(t)|)\right),$$

and $\omega(\Delta_n; f) \to 0$ for $\Delta_n \to 0$ for any continuous function f.   □

If meshes are not quasi-uniform, then we should require more of a function f than its continuity.

Prove that if $f \in C^{(j)}[a, b]$ for some $1 \leq j \leq 4$, then

$$\|f - S_n\|_{C[a, b]} = \mathcal{O}(\Delta_n^j).$$

14.6 B-splines

Suppose a mesh $x_0 < \ldots < x_n$ is embedded in an infinite simple mesh

$$M_\infty = \{\ldots x_{-1} < x_0 < \ldots < x_n < x_{n+1} < \ldots\}.$$

Then any spline of degree m can be represented by a linear combination of some basic m-degree splines zeroed outside intervals of the form $[x_{k-m}, x_k]$. If the mesh is quasi-uniform then all the basic splines can be obtained by changing x onto $x - x_k$ from *one* basic function.

Let us assume that the functions are defined on the whole real axis. If $f(x) = 0$ for all $x \notin [a, b]$, then $f(x)$ is said to be a function *with a finite support*. A support of a function is defined as the closure of its nonzero values. Notation: supp f.

Let M_∞ comprise all integers. By recurrence, construct the following functions:

$$B_0(x) = \begin{cases} 1, & 0 \le x \le 1, \\ 0, & \text{otherwise,} \end{cases}$$

$$B_m(x) = \int_0^1 B_{m-1}(x - y) B_0(y) \, dx, \quad m = 1, 2, \dots.$$

The functions B_m are called *B-splines*. Here are their main properties:

(1) B_m is a spline of degree m on the mesh M_∞.

(2) $B_m \in C^m$, and $B_m'(x) = B_{m-1}(x)$.

(3) supp $B_m = [0, m]$.

(4) Any spline of degree m om the mesh $0 < 1 < \dots < n$ is uniquely represented by a linear combination of the splines $B_m(x - k)$, where k is an integer running from $-m + 1$ to $n + m - 1$.

Verify that the functions $B_m(h^{-1}x - k)$ for integer k are the basic splines on the uniform mesh with step h.

14.7 Quasi-Local property and banded matrices

A natural spline does not possess the locality property: if we change f_k onto $\tilde{f}_k$ in just one node x_k, then the values of the spline vary everywhere. However, the natural spline still possess the *quasilocal property*: under a perturbation of f_k, the spline's values vary very little outside a neighborhood of the point x_k.

A rather profound property of banded matrices is recognized behind the quasi-local property.

Theorem 14.7.1 *Suppose that $A = [a_{ij}]$ is a nonsingular banded matrix of order n with band semiwidth not greater than L, that is,*

$$a_{ij} = 0 \quad \text{for} \quad |i - j| \ge L.$$

Let A have a nonzero diagonal, and assume that

$$q \equiv \|(\operatorname{diag} A)^{-1} \text{ off } A\| < 1.$$

Then, for the entries $a_{ij}^{(-1)}$ of the inverse matrix A^{-1}, the following inequalities hold:

$$|a_{ij}^{(-1)}| \le c \, \|(\mathrm{diag}\, A)^{-1}\| \, \frac{q}{1-q} \, q^{\frac{|i-j|}{L}}, \quad i,j = 1, \dots, n,$$

where $c > 0$ depends only on what norm $\| \cdot \|$ is used and possibly on the order of A.

Proof. Set $F \equiv (\mathrm{diag}\, A)^{-1}$ off A, and consider the Neuman series

$$(\mathrm{diag}\, A)\, A^{-1} = I + F + F^2 \dots .$$

Note that F^k is a banded matrix with the band semiwidth not greater than $(k-1)\, L$. It follows from an almost evident fact that the band semiwidth for the product AB does not exceed $L_1 + L_2 - 1$, where L_1 and L_2 are the bandwidths for A and B, respectively.

Fix i and j, and choose k so that

$$(k-1)\, L \le |i - j| < k\, L.$$

Then

$$g_{ij} \equiv \{\sum_{j=0}^{\infty} F^j\}_{ij} = \{\sum_{j=k}^{\infty} F^j\}_{ij},$$

and, hence, $|g_{ij}| \le \frac{c\, q^k}{1-q}$. $\square$

Corollary 14.7.1 *Let the natural splines $S(x)$ and $\tilde{S}(x)$ interpolate the values f_i and $\tilde{f}_i$ coinciding everywhere except for $i = k$. Let $\varepsilon \equiv |f_k - \tilde{f}_k|$. Then for some $c' > 0$,*

$$|S(x) - \tilde{S}(x)| \le c' \varepsilon \, \frac{\Delta_n^2}{\delta_n^2} \, \frac{1}{2^{|i-k|}}, \quad x \in [x_{i-1}, x_i].$$

Proof. Since T of the form (14.4.1) is a tridiagonal matrix, we have $L = 1$. Beside this, $q = \frac{1}{2}$ and $c = 1$ (for the norm $\| \cdot \|_{\infty}$).

If the value $f(x)$ gets different only at the node x_k, then in the right-hand side of the system $T\, u = \rho$, only three components may change: ρ_{k-1}, ρ_k, and ρ_{k+1}. Obviously, the perturbations are of the form $\mathcal{O}(\varepsilon/\delta_n)$ (as above, δ_n is the minimal step of the mesh).

For the components of the vectors $u = T^{-1}\rho$ and $\tilde{u} = T^{-1}\tilde{\rho}$,

$$|u_i - \tilde{u}_k| \le c_1 \frac{\varepsilon}{\delta_n^2} \, \frac{1}{2^{|i-k|}}, \quad c_1 > 0.$$

We must remember formula (14.4.2). $\square$

Exercises

1. Prove that for the matrix T defined by (14.4.1),

$$||T^{-1}||_\infty \leq \frac{1}{\min\limits_{1 \leq k \leq n-1}(h_k + h_{k+1})}.$$

2. Build the natural spline, given the nodes

$$x = -2, \quad -1, \quad 0, \quad 1, \quad 2$$

and the corresponding values

$$y = 0, \quad \frac{1}{6}, \quad \frac{2}{3}, \quad \frac{1}{6}, \quad 0.$$

3. Let a mesh $a = x_0 < \ldots < x_n = b$ be given, and assume that for a function f with period $b - a$, an interpolative cubic spline is to be found subject to the periodicity conditions

$$S'(x_0) = S'(x_0), \quad S''(x_n) = S''(x_n).$$

What does the algebraic system with respect to the quantities $u_k = S''(x_k)$ look like? Prove that it has a unique solution.

4. Given a mesh

$$a = x_0 < \ldots < x_n = b,$$

let us interpolate the values $f_0, \ldots, f_{n-1}, f_n = f_0$ by a cubic spline $S(x)$ satisfying the periodicity conditions. Prove that if a function $\phi \in C^2[-\infty, \infty]$ with period $b - a$ interpolates the same values, then

$$\int_a^b (\phi''(x))^2 \, dx \geq \int_a^b (S''(x))^2 \, dx.$$

5. Let

$$\phi \in C^r[a, b] \quad \text{and} \quad \phi(x_k) = f_k, \quad k = 0, 1, \ldots, n.$$

Prove that the minimum of a functional

$$E(\phi) \equiv \int_a^b (f^{(r)}(x))^2 \, dx$$

is attained on the function ϕ which is a spline of degree $2r - 1$ interpolating the values f_k and subject to the extra conditions

$$f^{(j)}(x_0) = f^{(j)}(x_n) = 0, \quad j = r, r+1, \ldots, 2r - 2.$$

6. The natural spline S interpolates the values of a function $f \in C^4 [a, b]$ on a mesh with maximal step h. Prove that

$$\|S^{(j)} - f^{(j)}\|_{C[a, b]} = O(h^{4-j}), \quad 0 \le j \le 3.$$

7. For the nodes $x = -2, -1, 0, 1, 2$, find the cubic spline $B(x) \in C^2$ subject to the conditions

$$B^{(r)} (\pm 2) = 0, \quad r = 0, 1, 2.$$

Is such a spline unique? Can it be an odd function?

8. Suppose that a mesh $a = x_0 < \ldots < x_n = b$ and the values $y_0, \ldots, y_n$ are given. Prove that if a function f minimizes the functional

$$J(f) \equiv \int_a^b (f'')^2 \, dx + \sum_{k=0}^n (f(x_k) - y_k)^2,$$

on $C^2 [a, b]$, then it must be a natural spline.

9. Prove that any spline of degree m on the mesh $0 < 1 < \ldots < n$ is uniquely represented by a linear combination of the splines $B_m (x - k)$ for integer k running from $-m + 1$ to $n + m - 1$.

10. A uniform mesh with step h does not include the point $x = 0$. Is it true that any spline of degree m on this mesh can be written as a linear combination of a finite number of functions $B_m (h^{-1} x - k)$ for integer k?

11. Given uniform meshes $\{x_k = kh, \quad k = 0,, \ldots, n\}$ on the interval $[0, b]$ with step $h = b/n$, assume that a function $f \in C^2$ is approximated by functions S_h of the form

$$S_h (x) = \sum_{k=-1}^{n+1} \alpha_k B_3 (h^{-1} x - k).$$

Let $\alpha_k = f(x_k)$ for $k = -1, 0, \ldots, n + 1$. Prove that

$$\|f - S_h\|_{C[0, b]} = O(h^2). \tag{$*$}$$

12. Prove that $(*)$ is still valid if $\alpha_k = f(x_k)$ only for $k = 0, \ldots, n$ whereas the values α_{-1} and α_{n+1} are determined from $f(x_0), f(x_1)$ and $f(x_n), f(x_{n-1})$, respectively, via a linear interpolation.

Is S_h an interpolative spline?

Lecture 15

15.1 Norm minimization

The theory and methods of minimizing a norm $||f - \phi||$ on the whole of a space of "simple" functions $\phi \in \Phi$ depend crucially on the norm type. One should distinguish between the two cases:

- *Uniform approximation.* $||f|| \equiv \sup_x |f(x)|$ (there can be no scalar product inducing such a norm).

- *Least squares method.* A norm in question is induced by some scalar product, for instance, $||f|| \equiv \left(\int_a^b |f(x)|^2 \, dx \right)^{\frac{1}{2}}$.

15.2 Uniform approximations

Let $f(x) \in C[a, b]$, and suppose that a polynomial $p_n(x)$ of degree not higher than n minimizes the norm $||f - p_n||_{C[a, b]}$. Such a polynomial is called *the best uniform approximant* for f.

In the theory of uniform approximations, the starring role is played by the notion of *the Chebyshev alternance points* for the function $R(x) = f(x) - p_n(x)$. The alternance points of degree m are the nodes of a simple mesh

$$a \leq x_1 \leq \ldots \leq x_m \leq b$$

possessing the following properties:

(1) $|R(x_i)| = \max_{a \leq x \leq b} |R(x)|, \quad i = 1, \ldots, m.$

(2) $R(x_i) R(x_{i+1}) < 0, \quad i = 1, \ldots, m-1.$

Denote the set of all such meshes on $[a, b]$ by $\mathcal{A}(m, a, b, R)$.

Theorem 15.2.1 (P. L. Chebyshev) *For a polynomial p_n of degree not higher than n to be the best uniform approximant for $f \in C[a, b]$, it is necessary and sufficient that $A(n + 2, a, b, R)$ is not empty.*

Proof of sufficiency. Assume that for some polynomial q_n of degree n,

$$\|f - q_n\| < \|f - p_n\|.$$

Then at the Chebyshev alternance points

$$|f(x_i) - q_n(x_i)| < |f(x_i) - p_n(x_i)| \quad \Rightarrow$$

the function $g(x) \equiv (f(x) - p_n(x)) - (f(x) - q_n(x))$ has the same sign at the points x_i as $R(x) = f(x) - p_n(x)$. Since the signs of $R(x_i)$ alternate, there is a zero of $g(x)$ inside every interval $[x_i, x_{i+1}]$. Hence, $g(x)$ has $n + 1$ zeroes on $[a, b]$. It could not be so if g is not identically zero, for $g(x)$ is a polynomial of degree n or less. $\square$

15.3   More on Chebyshev polynomials

Recall that the Chebyshev polynomials are defined as follows:

$$\begin{aligned}
T_0(t) &= 1, \quad T_1(t) = t, \\
T_{n+1}(t) &= 2t\,T_n(t) - T_{n-1}(t), \quad n = 1, 2 \dots.
\end{aligned}$$

We already know that

$$T_n(t) = \cos(n \arccos t) \quad \text{for} \quad |t| \leq 1. \tag{15.3.1}$$

Now we add to this that if $|t| \geq 1$, then

$$T_n(t) = \frac{1}{2}\left\{ \left(\frac{t + \sqrt{t^2 - 1}}{2}\right)^n + \left(\frac{t - \sqrt{t^2 - 1}}{2}\right)^n \right\}. \tag{15.3.2}$$

Indeed, the sequence $y_n \equiv T_n(t)$ satisfies the equation $y_{n+1} = 2t\,y_n - y_{n-1}$, $n = 1, 2, \dots$ for any initial values y_0, y_1. Consider a sequence of the form $y_n = z^n$, where $z \neq 0$, and require that it satisfy the same equation. The latter is equivalent to $z^2 - 2tz + 1 = 0$. Hence, $z = z_\pm$, where $z_\pm = \frac{t \pm \sqrt{t^2 - 1}}{2}$. $\Rightarrow$ For any c_+, c_- the sequence $y_n = c_- z_-^n + c_+ z_+^n$ satisfies the equation $y_{n+1} = 2t\,y_n - y_{n-1}$ for $n = 1, 2, \dots$. By way of getting the solution we are after, it remains to choose c_+, c_- to fulfill the initial conditions

$$\begin{aligned}
c_- \quad + \quad c_+ \quad &= \quad 1, \\
c_- z_- \quad + \quad c_+ z_+ \quad &= \quad t. \quad \square
\end{aligned}$$

An important property of Chebyshev polynomials: they are bounded in modulus by 1 on the interval $[-1, 1]$ uniformly in n and grow exponentially in n at any point outside this interval.

15.4 Polynomials of the least deviation from zero

A polynomial $q_n(x) = x^n + a_{n-1}x^{n-1} + \ldots + a_0$ is said to have *the least deviation from zero* on the interval $[a, b]$ if it has the least possible norm in $C[a, b]$ among all polynomials whose senior coefficient equals 1.

According to this definition,

$$\|q_n(x)\|_{C[a,\,b]} \leq \|x^n - p_{n-1}(x)\|_{C[a,\,b]}$$

for any polynomial $p_{n-1}(x)$ of degree $n-1$ or less. Therefore, the difference $x^n - q_n(x)$ is the best uniform approximant to the function x^n on the segment $[a, b]$ among all polynomials of degree $n-1$ or less.

It is easy to verify (do this) that the polynomial $2^{1-n}T_n(x)$ possesses $n+1$ Chebyshev alternance points on $[-1, 1]$. A polynomial $x^n - 2^{1-n}T_n(x)$ is of degree not higher than n, and on the strength of Chebyshev's theorem it is the best uniform approximant to x^n on $[-1, 1]$.

Consequently, the polynomial $2^{1-n}T_n(x)$ has the least deviation from zero on $[-1, 1]$.

In the case of an arbitrary segment $[a, b]$, consider the transformation $x = \frac{a+b}{2} + t\frac{b-a}{2}$ (it maps $[-1, 1]$ onto $[a, b]$) and the inverse:

$$t = \frac{2x - a - b}{b - a}.$$

Since the senior coefficient of $T_n(\frac{2x-a-b}{b-a})$ equals $\frac{2^{2n-1}}{(b-a)^n}$, we arrive at the following.

Theorem 15.4.1 *The polynomial*

$$Q_n(x) \equiv 2^{1-2n}(b-a)^n T_n\left(\frac{2x - a - b}{b - a}\right) \tag{15.4.1}$$

has the least deviation from zero on the segment $[a, b]$ and provides that

$$\|Q_n(x)\|_{C[a,\,b]} = 2^{1-2n}(b-a)^n. \tag{15.4.2}$$

15.5 The Taylor series and its discrete counterpart

A function $f(x) \in C^{n+1}[a, b]$ can be approximated by the n-term Taylor series at the point $c = \frac{a+b}{2}$:

$$P_n(x) = \sum_{k=0}^{n} \frac{f^{(k)}(c)}{k!}(x - c)^k.$$

It can also be approximated by the Lagrange polynomial $L_n(x)$ for the nodes $x_i = \frac{a+b}{2} + \frac{b-a}{2} t_i$, $i = 0, \ldots, n$, where t_i are the roots of the Chebyshev polynomial T_{n+1}.

For these two approximating devices, we have the following estimates:

$$\|f - P_n\|_{C[a,b]} \leq \frac{\|f^{(n+1)}\|_{C[a,b]}}{(n+1)!} \left(\frac{b-a}{2}\right)^{n+1}, \qquad (15.5.1)$$

$$\|f - L_n\|_{C[a,b]} \leq \frac{\|f^{(n+1)}\|_{C[a,b]}}{(n+1)!} \left(\frac{b-a}{2}\right)^{n+1} \frac{1}{2^{n-1}}. \quad (15.5.2)$$

One can see that the interpolative method using the Chebyshev nodes is capable of giving us far better accuracy.

15.6   Least squares method

Suppose that a norm is induced by a scalar product. In this case, the theory and methods of the best approximation on a subspace seem to be entirely transparent.

The theory dwindles down to the theorem that, for an arbitrary vector f and any (closed) subspace Φ, there is a uniquely determined expansion of the form

$$f = u + \phi, \quad \phi \in \Phi, \quad u \perp \Phi.$$

If $\Phi = \operatorname{span}\{v_1, \ldots, v_n\}$, then the best approximant $\phi = \alpha_1 v_1 + \ldots + \alpha_n v_n$ can be found by solving the linear system with the Gram coefficient matrix

$$\begin{bmatrix} (v_1, v_1) & \cdots & (v_n, v_1) \\ \cdots & \cdots & \cdots \\ (v_1, v_n) & \cdots & (v_n, v_n) \end{bmatrix} \begin{bmatrix} \alpha_1 \\ \cdots \\ \alpha_n \end{bmatrix} = \begin{bmatrix} (v_1, f) \\ \cdots \\ (v_n, f) \end{bmatrix}.$$

Best of all is to deal with an orthonormal basis of Φ. Then the Gram matrix turns to be the unity matrix. Remember that an orthonormal basis can be built through the Gram-Schmidt orthogonalization algorithm.

15.7   Orthogonal polynomials

Consider the space of all algebraic polynomials with the scalar product

$$(f, g) \equiv \int_a^b f(x) g(x) w(x)\, dx, \qquad (15.7.1)$$

where $w(x)$ is a continuous nonnegative function that is allowed to have only finite many zeroes on $[a, b]$, or, in the general case,

$$(f, g) \equiv \int_a^b f(x) g(x)\, dW(x), \qquad (15.7.2)$$

where $W(x)$ is a monotonically decreasing function and the integral is understood in the Stieltjes sense.

We still assume further on that the scalar product is of the form (15.7.1), but keep in mind that the results stand for the general case. The function $w(x)$ is said to be a *weight*, or weight function.

Any weight $w(x)$ generates a sequence of orthogonal polynomials q_n, $n = 0, 1, \ldots$, where the degree of q_n is equal to n. The orthogonal polynomials are determined up to a nonzero scalar multiplier.

It is remarkable that the orthogonal polynomials possess some common properties regardless of a concrete form of the weight.

15.8   Three-Term recurrence relations

Write

$$x\, q_n(x) = \sum_{k=0}^{n+1} a_{nk}\, q_k(x).$$

The polynomials are orthogonal, and, due to the integral formula for the scalar product, for $k \leq n - 2$,

$$a_{nk} = \frac{(x\, q_n,\, q_k)}{q_k,\, q_k} = \frac{(q_n,\, x\,,\, q_k)}{q_k,\, q_k} = 0.$$

We have obtained the *three-term recurrence relations* between the orthogonal polynomials:

$$x\, q_n(x) = a_{n\,n-1}\, q_{n-1}(x) + a_{nn}\, q_n(x) + a_{n\,n+1}\, q_{n+1}(x). \qquad (15.8.1)$$

Having normalized the polynomials so that $(q_n, q_n) = 1 \;\forall\; n$, we get

$$(x\, q_n,\, q_{n-1}) = (x\, q_{n-1},\, q_n) \quad \Rightarrow \quad a_{n\,n-1} = a_{n-1\,n} \quad \Rightarrow$$

$$x\, q_n(x) = b_{n-1}\, q_{n-1}(x) + a_n\, q_n(x) + b_n\, q_{n+1}(x), \qquad (15.8.2)$$

where

$$a_k \equiv a_{k\,k-1}, \quad b_k \equiv a_{kk}.$$

15.9   The roots of orthogonal polynomials

For $n \geq 1$, let us write

$$q_n(x) = (x - \zeta_1)\ldots(x - \zeta_m)\, p_{n-m}(x),$$

where $\zeta_1, \ldots, \zeta_m$ are pairwise distinct roots of the polynomial $q_n(x)$ located strictly inside the interval $[a, b]$ and having odd multiplicity. Let m be the greatest possible number of such roots. Then the polynomial $p_{n-m}(x)$ keeps the same sign for all $x \in [a, b]$.

If $m < n$, then, by the orthogonality of q_n to all the polynomials of smaller degree, we find that

$$\int_a^b (x - \zeta_1)^2 \ldots (x - \zeta_m)^2 \, p_{n-m}(x) \, w(x) \, dx = 0.$$

Such an equality is impossible. (Why?) Hence, $m = n$.

Conclusion: for $n \geq 1$ all the roots of q_n are real, pairwise distinct, and located strictly inside the interval $[a, b]$.

15.10   Three-Term relations and tridiagonal matrices

The three-term relations (15.8.2) can be expressed in the matrix-vector language as follows (check it):

$$x \, [q_0(x), \ldots, q_n(x)] = [q_0(x), \ldots, q_n(x)] \, T_n + a_n \, q_{n+1}(x), \quad (15.10.1)$$

where

$$T_n = \begin{bmatrix} a_0 & b_0 & & & & \\ b_0 & a_1 & b_1 & & & \\ & b_1 & a_2 & b_2 & & \\ & & \ldots & \ldots & \ldots & \\ & & & b_{n-2} & a_{n-1} & b_{n-1} \\ & & & & b_{n-1} & a_n \end{bmatrix}. \quad (15.10.2)$$

It is easy to see that $a_n \neq 0$ for all n. (Why?) If x is a root of the polynomial q_{n+1}, then it cannot be a root of q_n (otherwise, by the three-term relations x would have been a root for all the polynomials q_i for $i \leq n$, and in particular, for q_0 which is free of roots). The relation (15.8.1) acquires the form

$$[q_0(x), \ldots, q_n(x)] \, (T_n - x \, I) = 0,$$

and since $q_n(x) \neq 0$, we see that x is an eigenvalue of the matrix T_n.

Thus, every root of the polynomial q_{n+1} is an eigenvalue of the tridiagonal real symmetric matrix T_n of order $n + 1$.

This fact gives, at least, a sound algorithm for computing the roots of orthogonal polynomials (based on using the QR algorithm to find the eigenvalues of the tridiagonal matrices T_n).

Moreover, the relation with the tridiagonal matrices allows one to establish one more important property of the roots of orthogonal polynomials: the so-called interlacing inequalities.

15.11 Separation of the roots of orthogonal polynomials

For the roots $\lambda_1 > \ldots > \lambda_n$ and $\mu_1 > \ldots > \mu_{n-1}$ of q_n and q_{n-1}, respectively, the following *interlacing inequalities* hold:

$$\lambda_k > \mu_k > \lambda_{k+1}, \quad k = 1, \ldots, n-1. \tag{15.11.1}$$

This property inherent to orthogonal polynomials follows quite obviously from the interlacing properties for Hermitian matrices (see Lecture 5) and the results from Sections 15.9 and 15.10.

15.12 Orthogonal polynomials and the Cholesky decomposition

Write $q_i = q_{i0} + q_{i1} x + \ldots + q_{ii} x^i$ and consider a lower triangular matrix

$$L_n = \begin{bmatrix} q_{00} & & & \\ q_{10} & q_{11} & & \\ \ldots & \ldots & \ldots & \\ q_{n0} & q_{n1} & \ldots & q_{nn} \end{bmatrix}.$$

The orthogonality relations for the polynomials q_i and q_j for $0 \le i, j \le n$ can be expressed in the following way:

$$\int_a^b \left\{ L_n \begin{bmatrix} 1 \\ x \\ \ldots \\ x^n \end{bmatrix} [1 \; x \; x^2 \; \ldots \; x^n] L_n^T \right\}_{ij} w(x) \, dx = \begin{cases} 1, & i = j, \\ 0, & i \neq j. \end{cases}$$

Introduce the *matrix of moments*

$$M_n \equiv [(x^i, x^j)]_{ij=0}^n.$$

Then $L_n M_n L_n^T = I$, and hence,

$$M_n = L_n^{-1} L_n^{-T}.$$

Thus, the matrix L_n composed of the coefficients of orthogonal polynomials is the inverse to the lower triangular matrix from the Cholesky decomposition for the matrix of moments M_n.

Exercises

1. Prove the uniqueness of the best uniform approximation polynomial for $f \in C[a, b]$.

2. For any function $f \in C[a, b]$, prove the existence of the best uniform approximation polynomial.

3. Let a function $f \in C[-1, 1]$ be even. Prove that the best uniform approximation polynomial for f must be an even function. Must it be odd for an odd function?

4. Prove that the Chebyshev polynomials are the orthogonal polynomials on $[-1, 1]$ with weight $w(x) = 1/\sqrt{1 - x^2}$.

5. Find the coefficients of the three-term recurrence relations for the Legendre polynomials (by definition, these are the ones orthogonal on $[-1, 1]$ with weight $w(x) = 1$).

Lecture 16

16.1 Numerical integration

To obtain an integral of a function f numerically, one approximates f by a "simple" function ϕ and sets

$$I(f) \equiv \int_a^b f(x)\, dx \approx S(f) \equiv \int_a^b \phi(x)\, dx.$$

As "simple" functions, we usually choose those which are integrated analytically (for instance, polynomials or splines).

A vast class of numerical integration methods is described by the following *quadrature formula*:

$$S(f) = \sum_{i=1}^n d_i\, f(x_i). \tag{16.1.1}$$

A quadrature formula is defined by its nodes x_i and weights d_i. The weights are often represented in the form $d_i = \frac{b-a}{2} D_i$, where D_i does not depend on a and b.

16.2 Interpolative quadrature formulas

Consider the standard segment $[-1,\, 1]$ and map it onto $[a,\, b]$:

$$x = x(t) = \frac{a+b}{2} + \frac{b-a}{2}\, t.$$

Choose the nodes $t_1, \ldots, t_n \in [-1,\, 1]$ and set $x_i = x(t_i)$.

If the nodes are pairwise distinct, then we build the Lagrange polynomial

$$L_{n-1}(x) = \sum_{i=1}^n \prod_{\substack{j=1 \\ j \neq i}}^n f(x_i)\, \frac{x - x_j}{x_i - x_j}.$$

Having integrated it over $[a, b]$, we set

$$S(f) \equiv \int_a^b L_{n-1}(x)\, dx = \sum_{i=1}^n d_i\, f(x_i), \quad d_i = \frac{b-a}{2} \int_{-1}^1 \prod_{\substack{j=1 \\ j\neq i}}^n \frac{t - t_j}{t_i - t_j}\, dt.$$

$$(16.2.1)$$

Such quadrature formulas are sometimes referred to as the *Newton-Cotes for-mulas*.

Let $f \in C^n[a, b]$. Then using the error estimate for the Lagrange interpolation we obtain

$$|I(f) - S(f)| \leq \frac{\|f^n\|_{C[a,b]}}{n!} \left(\frac{b-a}{2}\right)^{n+1} \int_{-1}^1 |\prod_{j=1}^n (t - t_j)|\, dt. \quad (16.2.2)$$

16.3 Algebraic accuracy of a quadrature formula

If $I(f) = S(f)$ for all polynomials f of degree not higher than m and if $I(f) \neq S(f)$ for at least one polynomial of degree $m+1$, then the quadrature formula S is said to be of *algebraic accuracy m*.

Theorem 16.3.1 *A quadrature formula with n nodes is of algebraic accuracy $m \geq n - 1$ if and only if it is an interpolative quadrature formula of the form (16.2.1).*

Proof. The algebraic accuracy of the formula (16.2.1) is not less than $n - 1$. It is evident. If a formula of the form (16.1.1) is accurate for all polynomials of order $n - 1$, then, for d_i to be found, it is sufficient to apply it to the elementary Lagrange polynomials $l_i(x)$. □

16.4 Popular quadrature formulas

Set $h \equiv b - a$ and $M_m \equiv \|f^m\|_{C[a, b]}$.

The rectangular formula $(t_1 = 0)$: $S(f) = f(\frac{a+b}{2})\, h$. The error estimate (check it): $f \in C^1 \Rightarrow |I - S| \leq \frac{1}{4} M_1 h^2$.

It is curious that the same formula can be derived by integrating the Hermitian interpolating polynomial for the multiple node $t_1 = t_2 = 0$. For the standard segment, formally, $H_1(t) = f(0) + f'(0)\, t$, but, after integration, the term with the derivative disappears – thanks to the oddness. Now, we arrive at the following error estimate (check it):

$$f \in C^2 \Rightarrow |I - S| \leq \frac{1}{24} M_2 h^3.$$

The trapezoid formula ($t_1 = -1$, $t_2 = 1$): $S(f) = \frac{1}{2}(f(a) + f(b))h$. The error estimate (check it): $f \in C^2 \;\Rightarrow\; |I - S| \le \frac{1}{12} M_2 h^3$.

Simpson's formula: $t_1 = -1$, $t_2 = 1$, $t_3 = t_4 = 0$. On the standard segment, the Hermitian polynomial takes the form

$$H_3(t) = f(-1) + f(-1; 1)(t+1) \quad + \quad f(-1; 1; 0)(t+1)(t-1)$$
$$+ \quad f(-1; 1; 0; 0)(t+1)(t-1)t.$$

Due to the oddness, the last term vanishes after integration.

Complete the construction: find the weights and error estimate for Simpson's formula.

16.5 Gauss formulas

Given a number of nodes n, let us try to seek the quadrature formulas of the form (16.1.1) with maximal possible algebraic accuracy m. Such formulas are known as the *Gauss formulas*.

Theorem 16.5.1 *For any prescribed number of nodes n, the Gauss quadrature formula exists, is unique, and of algebraic accuracy $2n - 1$.*

Proof. Set

$$\omega_n(x) = \prod_{j=1}^{n}(x - x_j).$$

If there would be a formula of algebraic accuracy $2n$, then

$$I(\omega_n^2) = S(\omega_n^2) = 0,$$

which is impossible. Hence, $m \le 2n - 1$.

Assume that the formula (16.1.1) has algebraic accuracy $m = 2n-1$. Then

$$I(\omega_n(x) r_{n-1}(x)) = S(\omega_n(x) r_{n-1}(x)) = 0$$

for any polynomial $r_{n-1}(x)$ of degree not higher than $n - 1$. Therefore, the polynomial $\omega_n(x)$ is the nth polynomial of the sequence of the orthogonal polynomials on $[a, b]$ with weight 1. We are already aware that such a polynomial is determined uniquely up to a normalization. We also know that it must have n pairwise distinct roots inside $[a, b]$. We use these roots as the nodes x_i. By Theorem 16.3.1, the quadrature formula in question should be interpolative. Consequently, it is of the form (16.2.1).

Prove that the formula obtained has, indeed, algebraic accuracy $m = 2n-1$. Take an arbitrary polynomial $p_{2n-1}(x)$, and divide it with a remainder by $\omega_n(x)$:

$$p_{2n-1}(x) = q_{n-1}(x) \omega_n(x) + r_{n-1}(x).$$

From the linearity, orthogonality, and due to the appearance of the formula obtained, we infer that

$$
\begin{aligned}
I\left(p_{2n-1}\right) &= I\left(q_{n-1}\,\omega_n\right) + I\left(r_{n-1}\right) = I\left(r_{n-1}\right) \\
&= S\left(r_{n-1}\right) = S\left(q_{n-1}\,\omega_n\right) + S\left(r_{n-1}\right) = S\left(p_{2n-1}\right). \quad \square
\end{aligned}
$$

16.6   Compound quadrature formulas

The quadrature formulas considered above provide acceptable accuracy for small $h = b - a$. In the general case, one applies the *compound* quadrature formulas: subdivide the whole segment into some elementary segments, use some "elementary" quadrature formula on each of them, and then sum the results.

The error estimates for the compound quadrature formulas stem easily from those for the "elementary" quadrature formulas. For example, if $[a,\,b]$ is cut into subsegments of the same length h and the trapezoid formula is used on every subsegment, then the error for this compound quadrature formula will be of the form $\mathcal{O}\left(\frac{b-a}{h}\,h^3\right) = \mathcal{O}\left(h^2\right)$.

16.7   Runge's rule for error estimation

When subdividing a segment into smaller elementary segments, it is important to take into account the behavior of the function to be integrated. In case we know nothing beforehand, we can choose the elementary segments "step by step" progressing, for example, from the left to the right. If the current step size is h, then prior to integration, we should estimate the error expected and decide whether to accept, decrease, or increase the step size.

A "naive" approach to error estimation consists in using two quadrature formulas S_1 and S_2 and watching the value of $|S_1 - S_2|$ to bet on the accuracy. It might be advisable sometimes to judge the accuracy in a more elaborate way.

Assume that, for the step size h, we use a quadrature formula S_1 which is accurate for all polynomials of degree $n - 1$ or less. Expand the function $f(x)$ in the Taylor series at the middle point c of the current segment. Then

$$
I(f) - S_1(f) = \alpha\, f^{(n)}(c)\, h^{n+1} + \mathcal{O}\left(h^{n+2}\right).
$$

Denote by S_2 the quadrature formula obtained by applying S_1 to the two halves of the length h segment. Then, for the same constant α,

$$
I(f) - S_2(f) = \alpha\, f^{(n)}(c)\, \frac{h^{n+1}}{2^n} + \mathcal{O}\left(h^{n+2}\right).
$$

Prove this! Consequently, up to the $\mathcal{O}\left(h^{n+2}\right)$ terms, we obtain the following *Runge rule*:

$$I\left(f\right) - S_2\left(f\right) \approx \frac{S_2 - S_1}{2^n - 1}. \tag{16.7.1}$$

If we want the integral within the accuracy ε, then every next step should be done to provide that

$$\frac{|S_1 - S_2|}{2^n - 1} \leq \frac{h}{b - a}\varepsilon.$$

16.8 How to integrate "bad" functions

The quadrature formulas considered above are still not good enough if a function is not sufficiently smooth. Even in those cases, where the automatic adjustment of the step size is capable of solving the problem, it might take too many computations. There are two basic approaches to the numerical integration of "bad" functions.

- Split the function: $f = w + g$, where w is still a "bad" but prospectively "simple" function; integrate w individually (best of all, analytically) while using quadratures for a smoother function g.

- Factorize the function: $f = wg$, where w is made out to be a fixed "bad" function. A function g is assumed to be sufficiently smooth. It can be approximated by a polynomial p (for example, the interpolating polynomial), and then we are to find a sufficiently simple and accurate way of computing integrals from functions of the form $w\,p$. For example, this approach is advocated when $w = \sin(\omega x)$, allowing one to avoid using terribly many nodes to integrate a highly oscillating function. If w is a sign-preserving function, it can be regarded as a weight, and hence, the Gauss formulas can work.

Of course, other recipes exist. For instance, if a function has a singularity of the form x^α, where $0 < \alpha < \frac{1}{2}$, then a smoother function can be obtained easily by the change of variable $x = y^m$, where $m \geq 2$.

Exercises

1. Suppose thet a sequence of quadrature formulas is given:

$$S_n\left(f\right) = \sum_{i=1}^{n} d_{ni}\, f\left(x_{ni}\right), \quad x_{ni} \in [a, b].$$

Prove that if

$$\sum_{i=1}^{n} |d_{ni}| \to \infty \quad \text{for} \quad n \to \infty,$$

then there exists a function $f \in C[a, b]$ such that $S_n(f)$ does not converge to the integral of f over $[a, b]$.

2. Let a Newton-Cotes formula with an odd number of nodes n be applied to compute the integrals over the segment $[a, b]$ of the length $h = b - a$ for the functions $f \in C^{n+1}[a, b]$. Prove that the error in modulus does not exceed $c\|f^{n+1}\|_{C[a, b]} h^{n+2}$, where $c > 0$ is independent of f and h.

3. Prove that the weights in the Gauss formula are positive.

4. Let the Gauss formula with n nodes be applied to $f \in C^{2n}[a, b]$, $h = b - a$. Prove that the error in modulus does not exceed $c\|f^{2n}\|_{C[a, b]} h^{2n+1}$, where $c > 0$ is independent of f and h.

5. Estimate the error for the compound quadrature formula using Simpson's formula on every elementary segment of the length h.

6. Consider on $[a, b]$ a sequence of the compound trapezoid rules S_n with step size $h = (b - a)/n$. Prove that if $f \in C^4[a, b]$, then

$$\int_a^b f(x)\, dx = S_n(f) - \frac{1}{12}(f''(b) - f''(a))\, h^2 + \mathcal{O}(h^4).$$

Lecture 17

17.1 Nonlinear equations

Given a nonlinear equation $f(x) = 0$, we might wish to compute an *isolated* root z (separated from other roots). To begin with, it seems important to make out a range D (the smaller the better) localizing z.

In the most trivial fashion, the problem can be tackled by the following *bisection method*. If f is a continuous function then z belongs to any interval D through which the function alters the sign. We bisect D, then go back to a half-length interval, and proceed the same way.

The bisection method works even without a good enough initial guess. If a good quess is available, it makes sense (for smooth functions) to switch to some more rapid iterative methods.

17.2 When to quit?

We want to compute z within some accuracy ε and do not want to iterate very long. When do we to iterations?

The smallness of $f(x_k)$ is a very doubtful stopping criterion. (Why?) If we can go in for derivatives, then it seems sounder to quit when ($y = x_k$ or $y \approx x_k$)

$$|f(x_k)/f'(y)| \leq \varepsilon. \qquad (*)$$

Prove that if the derivative is continuous and the inequality $(*)$ is fulfilled for all $y \in [x_k - \varepsilon, x_k + \varepsilon]$, then $|z - x_k| \leq \varepsilon$.

Just the same, the criterion $(*)$ with $y = x_k$ is not error-proof. Let $g(t) = f(\alpha t)$. Then, for $t_k = x_k/\alpha$, the ratio $|g(t_k)/g'(t_k)|$ can be made arbitrarily small by the choice of α (regardless how close t_k is to a root of $g(t) = 0$).

17.3   Simple iteration method

Rewrite the equation $f(x) = 0$ as $x = F(x)$ (for example, set $F(x) = x - f(x)$), then choose an initial guess x_0 and consider the following *simple iteration method*:

$$x_{k+1} = F(x_k), \quad k = 0, 1, \dots .$$

A point z such that $z = F(z)$ is called a *fixed point* of the mapping F.

Theorem 17.3.1 *Let M be a complete metric space with distance ρ, and let a mapping $F : M \to M$ be contractive, in the sense that, for some $0 < q < 1$,*

$$\rho(F(x), F(y)) \leq q\rho(x, y) \quad \forall x, y \in M. \tag{17.3.1}$$

Then the equation $x = F(x)$ has a unique solution z, and, for any initial guess x_0, the simple iteration method converges to z with the speed of geometrical progression:

$$\rho(x_k, z) \leq \frac{q^k}{1-q}\rho(x_1, x_0). \tag{17.3.2}$$

Proof. For $m \geq k$, we find that

$$\rho(x_m, x_k) \leq \sum_{i=k}^{m-1} \rho(x_{i+1}, x_i) \leq \sum_{i=k}^{m-1} q^i \rho(x_1, x_0) \leq \frac{q^k}{1-q}\rho(x_1, x_0).$$

$\Rightarrow$ x_k is a Cauchy sequence $\Rightarrow$ since M is complete, x_k converges to some $z \in M$. It is clear that $F(z) = z$. (Why?) By transition to the limit, as $m \to \infty$, we obtain (17.3.2). $\square$

17.4   Convergence and divergence of the simple iteration

Let $F \in C^1[z-\delta, z+\delta]$, where z is a single fixed point for F. If $|F'(z)| < 1$, then, for some $\delta > 0$,

$$q \equiv \max_{|x-z|\leq\delta} |F'(x)| < 1 \quad \Rightarrow \quad |F(x)-F(y)| \leq q|x-y| \ \forall \ x, y \in [z-\delta, z+\delta].$$

In this case, F is a contractive mapping on the complete metric space $M = [z-\delta, z+\delta]$. Hence, the simple iteration method is convergent for any initial guess $x_0 \in M$.

If $|F'(z)| > 1$, then the simple iteration diverges for any initial guess $x_0 \neq z$. (Prove this!)

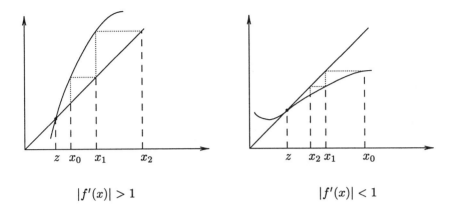

$$|f'(x)| > 1 \qquad\qquad\qquad |f'(x)| < 1$$

It is interesting to get on to analogous statements in the multidimensional case.

17.5   Convergence and the Jacobi matrix

Let $F : \mathbb{R}^n \rightarrow \mathbb{R}^n$ be a continuously differentiable mapping in a neighborhood of a single fixed point $z = F(z)$:

$$F(x) = [f_1(x), \ldots, f_n(x)]^T, \quad x = [x_1, \ldots, x_n]^T.$$

A matrix of the form

$$F'(x) = \begin{bmatrix} \dfrac{\partial f_1(x)}{\partial x_1} & \cdots & \dfrac{\partial f_1(x)}{\partial x_n} \\ \cdots & \cdots & \cdots \\ \dfrac{\partial f_n(x)}{\partial x_1} & \cdots & \dfrac{\partial f_n(x)}{\partial x_n} \end{bmatrix}$$

is called the *Jacobi matrix* of the mapping F at the point x. The continuous differentiability of F at the point x means that the Jacobi matrix entries (partial derivatives) are continuous functions at x.

Theorem 17.5.1 *Assume that a mapping $F : \mathbb{R}^n \rightarrow \mathbb{R}^n$ has a single fixed point $z = F(z)$ and is continuously differentiable in some neighborhood. If the spectral radius of the Jacobi matrix $F'(z)$ is less than 1, then for any initial guess x_0 in some neighborhood of z, the simple iteration method converges to z.*

The proof is similar to that for the one-dimensional case.

17.6   Optimization of the simple iteration

Usually we can pass from an equation $f(x) = 0$ to an equivalent one of the form $x = F(x)$ in many ways. For example,

$$F(x) = x - \alpha(x) f(x), \quad \text{where} \quad g(x) \neq 0 \; \forall \; x.$$

In particular, α can be an arbitrary nonzero constant. If z is an isolated root in search and $f'(z) \neq 0$, then α can be always chosen so that

$$|F'(z)| = |1 - \alpha f'(z)| < 1.$$

To accelerate the convergence, we need to diminish the value of $|F'(z)|$. (Why?) The best, but impractical, choice, of course, is $\alpha = 1/f'(z)$. However, we can also settle for some close value

$$\alpha = \frac{1}{f'(x_k)} \approx \frac{1}{f'(z)}.$$

In the end, the *Newton method* arises:

$$x_{k+1} = x_k - \frac{f(x_k)}{f'(x_k)}. \tag{17.6.1}$$

It is nothing but the simple iteration method for a function

$$F(x) = x - \frac{f(x)}{f'(x)}.$$

17.7   Method of Newton and Hermitian interpolation

The method of Newton is also known as the tangent method, the name stemming from its geometrical interpretation. We can also approach it using the interpolation ideas.

If we have x_k, let us set x_{k+1} to be the single root of the Hermitian interpolating polynomial $H(x) = f(x_k) + f'(x_k)(x - x_k)$. As is readily seen, this root is given by the formula (17.6.1).

Assume that

$$f \in C^2 \quad \text{and} \quad f'(z) \neq 0, \tag{17.7.1}$$

and consider the following two equalities:

$$f(z) - H(z) = \frac{f''(\xi_k)}{2}(z - x_k)^2 \quad \text{(the Hermitian interpolation error),}$$
$$H(x_{k+1}) - H(z) = f'(x_k)(x_{k+1} - z) \quad \text{(the Lagrange identity).}$$

Since $f(z) = H(x_{k+1}) = 0$, it follows that

$$e_{k+1} = -\frac{f''(\xi_k)}{2 f'(x_k)} e_k^2, \quad e_k \equiv z - x_k. \tag{17.7.2}$$

17.8 Convergence of the Newton method

We have everything to make an important resume right now: if f satisfies the conditions (17.7.1) and the Newton method converges for f, it converges *quadratically*.

By definition, a sequence x_k converges to z with order p if

$$\limsup_{k \to \infty} \left| \frac{e_{k+1}}{e_k^p} \right| \leq c < +\infty.$$

If $p = 1$, then the convergence is called linear. If $p > 1$, then it is superlinear, and if $p = 2$, then quadratic.

The condition $f'(z) \neq 0$ means that z is a *simple root* (the root of multiplicity 1). In the general case, z is said to be a root of multiplicity m if $f^{(j)}(z) = 0$ for $0 \leq j \leq m - 1$ while $f^{(m)}(z) \neq 0$.

The Newton method can also converge for a multiple root, but the convergence is no longer bound to be quadratic. For example, for $f(x) = x^2$, $e_{k+1} = e_k/2$, that is, the convergence is linear.

Theorem 17.8.1 *Let z be a simple root of an equation $f(x) = 0$, and assume that*

$$f \in C^2 [z - \delta, z + \delta], \qquad f'(x) \neq 0 \quad for \quad x \in [z - \delta, z + \delta],$$

$$\gamma \equiv \max_{|x-z| \leq \delta} |f''(x)| / \min_{|x-z| \leq \delta} |f'(x)| \neq 0.$$

Fix any $0 < \varepsilon < \min\{\delta, \gamma^{-1}\}$. Then the Newton method is convergent for any initial guess $x_0 \in [z - \varepsilon, z + \varepsilon]$, and, for all k, the following inequalities hold:

(a) $|e_{k+1}| \leq \gamma |e_k|^2$, *and* (b) $|e_k| \leq \gamma^{-1} (\gamma |e_0|)^{2^2}$.

Proof. Let $x_k \in [z - \varepsilon, z + \varepsilon]$. Then, from (17.7.2) and the definition of γ, we obtain

$$|e_{k+1}| \leq \gamma |e_k|^2 \leq (\gamma \varepsilon) \varepsilon \leq \varepsilon \quad \Rightarrow \quad x_{k+1} \in [z - \varepsilon, z + \varepsilon].$$

Thus, (a) takes place for all k. Multiply both sides of (a) by γ, and set $d_k \equiv \gamma |e_{k+1}|$. Thus, we come to

$$d_{k+1} \leq d_k^2 \quad \Rightarrow \quad d_k \leq d_0^{2^k}.$$

In line with what is required of the initial guess, $d_0 < 1$. Therefore, $e_k \to 0$. $\square$

Corollary. *Under the provisions of Theorem 17.8.1,*

$$\lim_{k \to \infty} \frac{e_{k+1}}{e_k^2} = -\frac{f''(z)}{2 f'(z)}.$$

17.9   Newton everywhere

The Newton method is exploited more frequently than one might imagine. It helps to divide numbers on a computer.

The division $c = a/b$ is usually performed through the two steps:

$$(1) \quad z = 1/b, \quad \text{and} \quad (2) \quad c = a \cdot z.$$

And z is computed by the Newton method – as a root of the equation

$$\frac{1}{x\,a} - 1 = 0.$$

It is clever to use the equation in this very form, for the Newton iterations turn out to be free of division:

$$x_{k+1} = x_k - \frac{(\frac{1}{x_k\,a} - 1)}{-\frac{1}{a\,x_k^2}} = 2\,x_k - a\,x_k^2.$$

17.10   Generalization for n dimensions

The Newton method is easy to generalize for solving a system of nonlinear equations of the form

$$\begin{cases} f_1(x_1, \ldots, x_n) = 0, \\ \qquad \cdots \\ f_n(x_1, \ldots, x_n) = 0. \end{cases} \quad\Longleftrightarrow\quad \begin{aligned} f(x) &= 0, \\ x, \; f(x) &\in \mathbb{R}^n, \\ f : \mathbb{R}^n &\to \mathbb{R}^n. \end{aligned}$$

In this case, the quantity $1/f'(x_k)$ is replaced by the inverse to the Jacobi matrix for the mapping f at the point x_k:

$$x_{k+1} = x_k - [f'(x_k)]^{-1} f(x_k). \tag{17.10.1}$$

Theorem 17.10.1 *Let z solve the equation $f(x) = 0$, and assume that, in the closed ball $B(\delta) \equiv \{x : \|x - z\|_\infty \le \delta\}$, the Jacobi matrix for the mapping $f : \mathbb{R}^n \to \mathbb{R}^n$ exists, is nonsingular, and subject to the following Lipschitz condition:*

$$\|f'(x) - f'(y)\|_\infty \le c\,\|x - y\|_\infty \quad \forall x, y \in B(\delta), \qquad c > 0.$$

Set

$$\gamma \equiv c \max_{\|z - x\|_\infty \le \delta} \|[f'(x)]^{-1}\|_\infty \quad \text{and} \quad 0 < \varepsilon < \min\{\delta, \gamma^{-1}\}.$$

Then, for any initial guess

$$x_0 \in B(\varepsilon) \equiv \{x : \|x - z\|_\infty \le \varepsilon\},$$

the Newton method is convergent, and the error vectors $e_k \equiv z - x_k$ satisfy the following inequalities:

$$(a) \quad \|e_{k+1}\|_\infty \le \gamma\,\|e_k\|_\infty^2, \quad \text{and} \quad (b) \quad \|e_k\|_\infty \le \gamma^{-1}\,(\gamma\,\|e_0\|_\infty)^{2^2}.$$

Proof. If the Jacobi matrix is continuous on the segment connecting x, $y \in \mathbb{R}^n$, then, by the Lagrange identity, there are points $\xi_1, \ldots, \xi_n$ on this segment such that

$$f(x) - f(y) = J_k(x - y),$$

where

$$J_k = \begin{bmatrix} \frac{\partial f_1(\xi_1)}{\partial x_1} & \cdots & \frac{\partial f_n(\xi_1)}{\partial x_n} \\ \cdots & \cdots & \cdots \\ \frac{\partial f_1(\xi_n)}{\partial x_1} & \cdots & \frac{\partial f_n(\xi_n)}{\partial x_n} \end{bmatrix}. \quad \Rightarrow$$

$$\begin{aligned} e_{k+1} &= e_k - [f'(x_k)]^{-1}(f(z) - f(x_k)) \\ &= e_k - [f'(x_k)]^{-1} J_k e_k \\ &= [f'(x_k)]^{-1}(f'(x_k) - J_k) e_k \quad \Rightarrow (a). \end{aligned}$$

From the Lipschitz condition,

$$\|f'(x_k) - J_k\|_\infty \le c \max_{1 \le j \le n} \|x_k - \xi_j\|_\infty \le c\|x_k - z\|_\infty.$$

Hence, if $\|z - x_k\|_\infty \le \varepsilon$, then $\|z - x_{k+1}\|_\infty \le (\gamma\varepsilon)\varepsilon \le \varepsilon$. Since (a) is valid for every k, we obtain (b) immediately. $\square$

17.11 Forward and backward interpolation

The interpolative approach to the Newton method is also interesting in that it suggests a general method of constructing iterative algorithms.

Forward interpolation. Having the points x_k, $x_{k-1}, \ldots, x_{k-m}$ and the corresponding values of a function f, we build the interpolative polynomial $L(x)$ of degree m and take as x_{k+1} one of its roots. In the case of pairwise distinct points, we deal with the Lagrange polynomial, and, otherwise, with the Hermitian polynomial.

To make this idea a success, we need know how to select between the roots of $L(x)$.

Backward interpolation. Having the points x_k, $x_{k-1}, \ldots, x_{k-m}$ and the corresponding values y_k, $y_{k-1}, \ldots, y_{k-m}$ of $f(x)$, we build the polynomial $P(y)$ of order m interpolating the values of the inverse function $f^{-1}(y)$ (these are x_k, $x_{k-1}, \ldots, x_{k-m}$) at the points y_k, $y_{k-1}, \ldots, y_{k-m}$. Then, it is natural to set $x_{k+1} = P(0)$.

The Lagrange and Hermitian interpolation error estimates give a base for analyzing such methods. In principle, one might devise a method with any prescribed convergence order.

17.12   Secant method

The forward and backward Lagrange interpolating polynomials of order 1 lead to the same method known as the *secant method*.

Having different points x_{k-1} and x_k, we build the Lagrange polynomial

$$L(x) = f(x_{k-1}) \frac{x - x_k}{x_{k-1} - x_k} + f(x_k) \frac{x - x_{k-1}}{x_k - x_{k-1}}$$

and find its single root:

$$x_{k+1} = x_k - f(x_k) \frac{x_k - x_{k-1}}{f(x_k) - f(x_{k-1})}. \tag{17.12.1}$$

On the other hand, having different values $f(x_{k-1})$ and $f(x_k)$ and interpolating the inverse function, we find that

$$P(y) = x_{k-1} \frac{y - f(x_k)}{f(x_{k-1}) - f(x_k)} + x_k \frac{y - f(x_{k-1})}{f(x_k) - f(x_{k-1})}.$$

It is easy to verify that $P(0) = x_{k+1}$.

Suppose that z is the root in search and $e_k \equiv z - x_k$. Let $f \in C^2$. Then, for one step of the secant method, we obtain the following relations:

$$f(z) - L(z) = \frac{f''(\xi_k)}{2} (z - x_k)(z - x_{k-1}),$$
$$L(x_{k+1}) - L(z) = \frac{f(x_k) - f(x_{k-1})}{x_k - x_{k-1}} (x_{k+1} - z) = f'(\zeta_k)(x_{k+1} - z). \Rightarrow$$

$$e_{k+1} = -\frac{f''(\xi_k)}{2 f'(\zeta_k)} e_k e_{k-1}. \tag{17.12.2}$$

17.13   Which is better, Newton or secant?

Let $f \in C^2$, $f'(z) \neq 0$, and assume that the secant method is convergent. According to (17.12.2), for some $\gamma > 0$, we get on to

$$|e_{k+1}| \leq \gamma |e_k| |e_{k-1}|.$$

Introduce the quantities $d_k \equiv \gamma |e_k|$, and assume that

$$d_0 \leq d < 1, \quad d_1 \leq d < 1.$$

Then $d_2 \leq d_1 d_0 \leq d^2$, $d_3 \leq d_2 d_1 \leq d^5$, and so on. In the general case,

$$d_k \leq d^{\phi_k}, \tag{17.13.1}$$

where

$$\phi_0 = \phi_1 = 1; \quad \phi_k = \phi_{k-1} + \phi_{k-2}, \quad k = 2, 3, \ldots . \tag{17.13.2}$$

The quantities ϕ_k defined by the recurrence relations (17.13.2) are called *Fibonacci numbers*. Verify that

$$\phi_k = \frac{1}{\sqrt{5}} \left(\left(\frac{1+\sqrt{5}}{2} \right)^{k+1} - \left(\frac{1-\sqrt{5}}{2} \right)^{k+1} \right).$$

Therefore,

$$\phi_k = \mathcal{O} \left(\left(\frac{1+\sqrt{5}}{2} \right)^k \right).$$

The Newton method errors are upper bounded by d^{2^k}. Since

$$\frac{1+\sqrt{5}}{2} \approx 1.618 < 2,$$

we infer that the Newton method converges faster. However, every iteration of it is twice heavier: we compute the function and derivative values. Consequently, from the standpoint of the total cost, the secant outperforms the Newton. A more detailed comparison of the methods can be found in the nice book of A. M. Ostrowski.[1]

Exercises

1. Let $F \in C^1 [z - \delta, z + \delta]$, where z is a single fixed point for F. Can the simple iteration method converge to z when $|F'(z)| = 1$? Can it diverge in this case?

2. Assume that a mapping $F : \mathbb{R}^n \to \mathbb{R}^n$ has a unique fixed point $z = F(z)$ and is continuously differentiable in a vicinity of z. Prove that if every eigenvalue of the Jacobi matrix for F at z is greater in modulus than 1, then the simple iteration method is divergent.

3. A mapping $F : \mathbb{R}^n \to \mathbb{R}^n$ has a unique fixed point $z = F(z)$ and is continuously differentiable in a vicinity of z. There is at least one eigenvalue of $F'(z)$ which is greater in modulus than 1. Can the simple iteration converge for any initial guess x_0 sufficiently close to z?

4. Clarify the convergence of the simple iteration for the following equations:

$$(1) \quad x = e^{2x} - 1; \quad (2) \quad x + \ln x = \frac{1}{2}; \quad (3) \quad x = \operatorname{tg} x.$$

---

[1] A. M. Ostrowski. *Solution of Equations and Systems of Equations.* University of Basel, Academic Press, New York and London, 1960.

5. Check that $z = [1, 1, 1]^T$ is one of the solutions of the equation $f(x) = 0$, where $f : \mathbb{R}^3 \rightarrow \mathbb{R}^3$ is of the form

$$f(x) = \begin{bmatrix} x_1 x_2^3 + x_2 x_3 - x_1^4 - 1 \\ x_2 + x_2^2 + x_3 - 3 \\ x_2 x_3 - 1 \end{bmatrix}.$$

Will the Newton method converge to z for all initial guesses sufficiently close to z ?

6. People say that the Newton method was tested for the first time on the equation

$$f(x) \equiv x^5 - 2x - 5 = 0.$$

Take $x_0 = 2$ and carry out two Newton iterations. Prove that the equation possesses a single real root z and that $|z - x_2| \leq 10^{-4}$.

7. Produce an example of an infinitely differentiable function f for which the equation $f(x) = 0$ has a root z such that the Newton method does not converge to z for any $x_0 \neq z$.

8. Suppose that the Newton method is applied to solve the equation $x^2 = a$ for $1 \leq a \leq 4$. The initial guess x_0 is chosen to be the value $p_1(a)$, where $p_1(t)$ is a polynomial of order 1 giving the best uniform approximation to the function $\sqrt{t}$ on $[1, 4]$. Find the coefficients of $p_1(t)$, and prove that $|x_4 - \sqrt{a}| \leq \frac{1}{2} 10^{-25}$.

9. A function $f \in C^{p+1}$ has an isolated zero z of multiplicity p. Consider the iterative process

$$x_{k+1} = x_k - p \frac{f(x_k)}{f'(x_k)},$$

and prove that if it converges to z, then it converges quadratically, and the errors $e_k \equiv z - x_k$ satisfy the limiting relation

$$\lim_{k \rightarrow \infty} \frac{e_{k+1}}{e_k^2} = \frac{f^{(p+1)}(z)}{p(p+1) f^{(p)}(z)}.$$

Lecture 18

18.1 Minimization methods

It is difficult (practically impossible) to invent a serious problem that would have not been reduced to a search for the minimum of a functional over some range. That is why the design and analysis of minimization methods is a vast field from which we select for discussion only some useful ideas and methods.

To begin with, note that the minimal value and the minimum point are not sought with the same accuracy. If the minimal value $f_{\min}$ of $f \in C^2$ is attained at the point $z_{\min}$, then

$$f\left(z_{\min} + \delta\right) \ = \ f\left(z_{\min}\right) \ + \ \mathcal{O}\left(||\delta||^2\right).$$

This implies that if $f_{\min}$ is computed within the accuracy ε, then $z_{\min}$ can be obtained, at best, with accuracy of the order $\sqrt{\varepsilon}$.

18.2 Newton again

Let $x \ = \ [x_1, \ldots, x_n]^T \in \mathbb{R}^n$, and let a functional $f(x) \in C^2$ have a unique minimizing point z. Then z satisfies the equation

$$f'(x) \equiv \operatorname{grad} f(x) \equiv \left[\begin{array}{c} \frac{\partial f(x)}{\partial x_1} \\ \ldots \\ \frac{\partial f(x)}{\partial x_n} \end{array} \right] = 0.$$

The equation $f'(x) \ = \ 0$ can be solved, for example, by the Newton method:

$$x_{k+1} \ = \ x_k - [f''(x_k)]^{-1} f'(x_k), \tag{18.2.1}$$

where $f''(x) \ = \ [\operatorname{grad} f(x)]'$ is the Jacobi matrix for the mapping $\operatorname{grad} f(x)$, sometimes called the *Hesse matrix*.

A convergence theory for the method (18.2.1) can be gotten from what we already know of the Newton method for solving nonlinear systems.

18.3   Relaxation

In the Newton method, unfortunately, one has to compute the second-order derivatives and to start from a sufficiently good initial guess. Methods exist that are free of either.

A search for the minimum is like a descent from a hill by moderate steps: at each point x_k, one needs to select a descent direction p_k and also a step size not far too large lest it lead to a nearby hill's slope. Set

$$x_{k+1} = x_k + \alpha_k\, p_k, \tag{18.3.1}$$

where p_k is a direction in which the functional $f \in C^1$ goes down:

$$(p_k, -f'(x_k)) \geq c\,\|p_k\|_2\,\|f'(x_k)\|_2, \quad c > 0, \tag{18.3.2}$$

and $\alpha \geq 0$ fulfills the following *relaxation* condition:

$$f(x_k + \alpha_k\, p_k) \leq f(x_k) - \tau\,\alpha_k\,(p_k, -f'(x_k)), \tag{18.3.3}$$

where $0 < \tau < 1$ is a *constant of relaxation*.

The condition (18.3.2) is bound up with the fact that the local fastest descent direction for a functional is that of its antigradient. It is met obviously for $p_k = -f'(x_k)$. In this case, the method is termed the *gradient method*.

18.4   Limiting the step size

If the condition (18.3.2) is fulfilled while $f \in C^1$ and $f'(x_k) \neq 0$, then the condition (18.3.3) is satisfied for all sufficiently small α. (Prove this!)

Consequently, an appropriate α can be always found by *limiting the step size*: try $\alpha = 1$, check the relaxation condition, in case it is not fulfilled, take $\alpha/2$, and so on but quit right away when the relaxation condition has become true. (The latter is important since it is no use to keep the step size *too small*, and also it permits a lower bound on the step size required in the convergence proof.)

Lemma 18.4.1 *Let $f \in C^2(\mathbb{R}^n)$ and $\|f''(x)\|_2 \leq M \; \forall\, x \in \mathbb{R}^n$, and assume that, at any point x_k, the descent direction p_k satisfies the condition (18.3.2) with the same $c > 0$. Then, at any point x, the relaxation condition with the constant τ is fulfilled for all $0 \leq \alpha_k \leq \hat{\alpha}$, where*

$$\hat{\alpha} = (1 - \tau)\,\frac{2\,c\,\|f'(x_k)\|}{M\,\|p_k\|}. \tag{18.4.1}$$

Proof. By the Taylor formula with the remainder term in Lagrange's form,

$$f(x_k + \alpha\, p_k) = f(x_k) + \alpha\,(f'(x_k),\, p_k) + \frac{\alpha^2}{2}\,(f''(\xi)\, p_k,\, p_k) \quad \Rightarrow$$

$$f(x_k) - f(x_k + \alpha p_k) \geq \tau \alpha(-f'(x_k), p_k)$$
$$+ \alpha(-f'(x_k), p_k)\{1 - \tau - \tfrac{\alpha M \|p_k\|}{2c \|f'(x_k)\|}\}.$$

With the expression in the curved brackets being nonnegative, we may put it aside to proceed to the relaxation condition. □

Corollary 18.4.1 *If α_k is chosen by limiting the step size, it is guaranteed that*

$$\alpha_k \geq (1 - \tau) \frac{c \|f'(x_k)\|}{M \|p_k\|}. \tag{18.4.2}$$

Theorem 18.4.1 *Suppose that the method (18.3.1) – (18.3.3) with relaxation by limiting the step size is applied to minimize a functional f bounded from below under the hypotheses of Lemma 18.4.1. Then, for an arbitrary initial guess $x_0 \in R^n$,*

$$\lim_{k \to \infty} \|f'(x_k)\|_2 = 0. \tag{18.4.3}$$

Proof. From the relaxation condition and Corollary 18.4.1,

$$f(x_k) - f(x_{k+1}) \geq \tau \alpha_k (-f'(x_k), p_k) \geq \frac{\tau(1-\tau)}{2} c^2 \|f'(x_k)\|^2.$$

The left-hand side tends to zero, since the sequence $f(x_k)$ is monotonically decreasing and bounded from below. □

The key question is how does the relative step size $\alpha \frac{\|p_k\|}{\|f'_k\|}$ behave. What the condition $\|f''\|_2 \leq M$ serves for is to provide a uniform lower bound for the step size.

If the sequence x_k (or any subsequence) converges to z, then, z is the minimizing point for f. (Why?)

18.5 Existence and uniqueness of the minimum point

Lemma 18.5.1 *Suppose that $f \in C^2(\mathbb{R}^n)$ and there exist positive constants m and M such that, for all $x, y \in \mathbb{R}^n$,*

$$m \|y\|^2 \leq (f''(x) y, y) \leq M \|y\|^2. \tag{18.5.1}$$

Then the minimum point z for f exists and is unique.

Proof. If f is not bounded from below, then there is a sequence of points y_k such that $f(y_k) \to -\infty$. This sequence y_k cannot be bounded. (Why?) Taking into account (18.5.1), we obtain

$$f(y_k) \geq f(y_1) - \|f'(y_1)\| \|y_k - y_1\| + \frac{m}{2} \|y_k - y_1\|^2. \tag{*}$$

It follows that $f(y_k) \to +\infty$, and the contradiction means that in reality f is bounded from below.

Due to (18.5.1), the set $\{x : f(x) \le f(x_0)\}$ is compact (prove this). $\Rightarrow$ The minimum point exists.

To prove uniqueness, assume that the points y_1 are y_k both the minimum points. Since $f'(y_1) = 0$, we deduce from $(*)$ that $\|y_k - y_1\| = 0$ $\Rightarrow$ $y_1 = y_k$. $\square$

18.6 Gradient method limiting the step size

For brevity, let us write $f_k \equiv f(x_k)$, $f'_k \equiv f'(x_k)$, and so on, and accept the notation

$$\varepsilon_k \equiv f_k - f(z), \quad e_k \equiv x_k - z,$$

where z is the minimum point for f.

In the gradient method, $p_k = -f_k$, and hence,

$$x_{k+1} = x_k - \alpha_k f'_k, \quad f_{k+1} \le f_k - \tau \alpha_k \|f'_k\|. \tag{18.6.1}$$

Theorem 18.6.1 *Under the hypotheses of Lemma (18.5.1), the gradient method limiting the stepsize is convergent for an arbitrary initial guess x_0 by geometric progression:*

$$\varepsilon_k \le q^k \varepsilon_0, \quad \|e_k\| = \mathcal{O}(q^{k/2}), \tag{18.6.2}$$

where $0 < q < 1$.

Proof. Using Corollary 18.4.1 for the case $p_k = -f'_k$, $\alpha_k \ge \alpha \equiv \frac{1-\tau}{M}$, and then, in chime with the relaxation condition,

$$\varepsilon_{k+1} \le \varepsilon_k - \tau \alpha \|f'_k\|^2.$$

To obtain an inequality of the form $\varepsilon_{k+1} \le q\varepsilon_k$, it is sufficient to make sure of

$$\|f'_k\|^2 \ge \gamma \varepsilon_k, \tag{18.6.3}$$

where $0 < q \equiv 1 - \tau \alpha \gamma < 1$.

It is easy to derive (18.6.3) with $\gamma = 2m^2/M$. Indeed,

$$(f'_k, e_k) = (f'_k - f'(z), e_k) = (f''(\xi) e_k, e_k) \quad \Rightarrow$$

$$m\|e_k\|^2 \le \|f'_k\| \|e_k\| \quad \Rightarrow \quad \|f'_k\|^2 \ge m^2 \|e_k\|^2 \ge \frac{2m^2}{M} \varepsilon_k.$$

A neater result asserts that (18.6.3) stands with $\gamma = \frac{m^2}{M} + m$. To this end, consider the Taylor series at z:

$$f(z) = f_k + (f'_k, -e_k) + \frac{1}{2}(f''(\zeta) e_k, e_k) \quad \Rightarrow$$

$$\varepsilon_k \leq \frac{||f_k'||^2}{m} - \frac{m}{2}||e_k||^2 \leq \frac{||f_k'||^2}{m} - \frac{m}{M}\varepsilon_k \quad \Rightarrow ||f_k'||^2 \geq (\frac{m^2}{M} + m)\varepsilon_k.$$

Thus, $\varepsilon_{k+1} \leq q\varepsilon_k$ for $q = 1 - \frac{\tau(1-\tau)}{M}(\frac{m^2}{M} + m)$. $\square$

Prove that $q \geq \frac{1}{2}$. This bound pertains exclusively to the claim that α_k is chosen by halving the step size.

18.7   Steepest descent method

The steepest descent method is the gradient method in which, on every step, α_k is chosen by minimizing $f(x_k - \alpha f_k')$ over all $\alpha \in \mathbb{R}$. Under the hypotheses of Lemma (18.5.1), it converges by geometric progression. (Prove this!)

Suppose the steepest descent method is applied to minimize the *quadratic functional*

$$f(x) \equiv \frac{1}{2}(Ax, x) - (b, x), \quad A = A^T \in \mathbb{R}^{n \times n}, \ b \in \mathbb{R}^n, \qquad (18.7.1)$$

where

$$0 < m||x||^2 \leq (Ax, x) \leq M||x||^2 \quad \forall x \neq 0. \qquad (18.7.2)$$

Proposition. *For the quadratic functional case,*

$$\varepsilon_{k+1} \leq \left(\frac{M-m}{M+m}\right)^2 \varepsilon_k. \qquad (18.7.3)$$

Proof. In this case, $f'(x) = Ax - b$ (check this), and the minimum point is $z = A^{-1}b$. It is easy to show also that

$$f(x) - f(z) = \frac{1}{2}(A(x - z), x - z).$$

Consider the gradient method with a constant step size α:

$$x_{k+1} = x_k - \alpha(Ax_k - b) \quad \Rightarrow \quad e_{k+1} = (I - \alpha A)e_k.$$

At the same time (check this),

$$\begin{aligned}
\varepsilon_{k+1} = \frac{1}{2}(Ae_{k+1}, e_{k+1}) &= \frac{1}{2}(A(I - \alpha A)e_k, (I - \alpha A)e_k) \\
&= \frac{1}{2}((I - \alpha A)Ae_k, (I - \alpha A)e_k) \\
&\leq ||I - \alpha A||_2^2 (Ae_k, e_k) = ||I - \alpha A||_2^2 \varepsilon_k.
\end{aligned}$$

One can show that (do this)

$$||I - \alpha A||_2 = \max\{|1 - \alpha m|, |1 - \alpha M|\}.$$

The right-hand side of the expression takes its minimal value when $\alpha = \frac{2}{M+m}$, and the value itself is $\frac{M-m}{M+m}$ (check this). For this very α, the errors of the gradient method with constant step size satisfy the inequality (18.7.3).

If the gradient method with step size α and the steepest descent method start from the same point x_k, then the latter is bound to provide an equal or smaller error ε_{k+1}. □

In a less trivial way[1] the inequality (18.7.3) was proved by L. V. Kantorovich in 1947.

18.8 Complexity of the simple computation

An obvious "drawback" of the gradient method, of course, is that it requires computing gradients. It seems that evaluation of grad f at one point in $\mathbb{R}^n$ should be at least n times more expensive than getting a value of f at one point. In the early 80s, W. Baur and V. Strassen[2] showed that this is not the case. Although the "price" of the gradient is higher than that of the functional, it is only finitely many times higher (independent of n)!

For a solid formulation, we need to define rigorously what we mean by the price of a computation.

Suppose that there is a reserve $\mathcal{O}$ of "elementary" binary operations, i.e., those of the form $w = a(u, v)$ (for example, $w = u + v$ or $w = uv$), and consider a sequence of operations as follows:

$$y_i = x_i, \quad 1 \leq i \leq n;$$

$$y_i = a_i(y_{i'}, y_{i''}), \quad n+1 \leq i \leq n+m,$$

where $1 \leq i' \leq i'' < i$ for all $n+1 \leq i \leq n+m$. We call such a sequence a *simple computation*, and the number m will be termed its *complexity* (price).

A simple computation can be regarded as an algorithm for computing any of the quantities y_{n+k}, $k = 1, \ldots, m$, or an ensemble of them.

If the elementary operations are of the form $a_i(u, v) = c_i u + d_i v$, where c_i, d_i are fixed constants, then the corresponding simple computation is referred to as a *linear computation*.

18.9 Quick computation of gradients

Let us require that $\mathcal{O}$ contain the operations $u \pm v$, uv and provide that, for any operation $a(u, v) \in \mathcal{O}$, the partial derivatives $a'(u, v) \equiv \frac{\partial a}{\partial u}(u, v)$ and

---

[1] D. K. Faddeev and V. N. Faddeeva. *Computational Methods of Linear Algebra*. San Francisco – London, 1963.

[2] W. Baur and V. Strassen. The complexity of partial derivatives. *Theor. Comput. Sci.* 22: 317–330 (1983).

$a''(u, v) \equiv \frac{\partial a}{\partial v}(u, v)$ can be found by simple computations of complexity not higher than c. Then the following holds.

Theorem 18.9.1 *Suppose that a functional f of n variables is defined by a simple computation of complexity m. Then, n components of the gradient of f and the value of f itself at one point can be defined by a common simple computation of complexity not higher than $(5 + 2c)m$.*

Proof. Introduce the quantities

$$u_{ij} \equiv \frac{\partial y_i}{\partial x_j},$$

and note that, for any fixed j, the quantities u_{ij} satisfy the following linear equations:

$$u_{1j} = \ldots = u_{j-1\,j} = 0,$$
$$u_{jj} = 1,$$
$$u_{j+1\,j} = \ldots = u_{nj} = 0;$$

$$-a_i'\, u_{i'\,j} - a_i''\, u_{i''\,j} + u_{ij} = 0, \quad n+1 \le i \le n+m.$$

In the matrix notation, it reduces to

$$VU = Z,$$

where

$$V = \begin{bmatrix} I_n & 0_{n\times m} \\ V_{21} & V_{22} \end{bmatrix} \in \mathbb{R}^{(n+m)\times(n+m)},$$

$$U = [u_{ij}] \in \mathbb{R}^{(n+m)\times n}, \quad Z = \begin{bmatrix} I_n \\ 0_{m\times n} \end{bmatrix} \in \mathbb{R}^{(n+m)\times n}.$$

The matrix V is lower triangular with units on the main diagonal; its first n rows coincide with those of the unity matrix; in every row i for $i > n$ there can be at most two nonzeroes, apart from units on the main diagonal.

Without loss of generality, we assume that $f(x) = y_{n+m}$. Then, obviously,

$$\mathrm{grad}\, f = [u_{n+m\,1}, \ldots, u_{n+m\,n}].$$

Of special interest to us are only the first n components of the last row of the matrix V^{-1}, or, in other words, the first n components of the solution to a linear system

$$[u_{n+m\,1} \ldots u_{n+m\,n}, \ldots]V = [0 \ldots 0\, 1].$$

This solution is computed by the backward substitution algorithm which involves, in this case, only multiplications and additions. The number of multiplications is equal to that of additions, and either coincides with the number of nonzeroes under the main diagonal of V. Beyond these operations, we need to count those that compute the quantities $u_{i'}$, $u_{i''}$, $a'_i(u_{i'}, u_{i''})$, and $a''(u_{i'}, u_{i''})$.   □

The theorem is easy to extend to elementary operations having p arguments, replacing $(5 + 2\,c)\,m$ by $(1 + (2 + c)\,p)\,m$. (Think this over.)

The proof of the theorem suggests a way of converting a code computing the functional into another code that simultaneously computes the functional and the gradient.

18.10   Useful ideas

Apparently, every minimization method has its own deficiency. In spite of the global convergence, the gradient methods might be very sluggish. For example, below is a picture the way the steepest descent method behaves when minimizing a quadratic functional in $\mathbb{R}^2$:

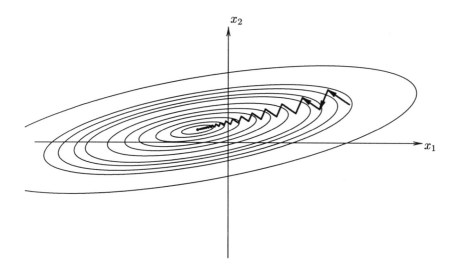

If the level lines for f are stretched too far, then the direction to the minimum point in search can be almost orthogonal to the gradient. One could try "more judicious" descent directions. For example, by the two points x_{k-1} and x_k, we can find a new point $y_k = x_k + (x_k - x_{k-1})\,\beta_k$ and, then, make a step in the antigradient direction computed at this point: $x_{k+1} = x_k - \alpha_k\,f'(y_k)$. This is the idea behind the so-called *ravine method*.

Another idea might be to seek x_{k+1} of the form

$$x_{k+1} = x_k + \alpha\, f'(x_k) + \beta\, (x_k - x_{k-1})$$

with α and β chosen to minimize $f\,(x_k + \alpha\, f'(x_k) + \beta\,(x_k - x_{k-1}))$. Instead of the one-dimensional minimization of the gradient methods we get on to the two-dimensional minimization at every step. This is the idea behind the *conjugate gradients method.*

One more idea is almost evident: to revamp the Newton method by replacing the second-order derivatives with some their approximations. One can also think of approximating the inverse to the Jacobi matrix by some simpler devices. This could evolve into *quasi-Newton methods.*

Finally, a familiar relaxation idea can make the convergence "more global." In particular, we can consider the following *damped Newton method*:

$$x_{k+1} = x_k - \alpha_k\, [f''(x_k)]^{-1}\, f'(x_k).$$

If α_k is chosen to satisfy a relaxation condition like (18.3.3) with constant $0 < \tau < 1/2$, then under the hypotheses of Lemma 18.5.1, one can prove that the method converges superlinearly for an arbitrary initial guess. Moreover, if the Hesse matrix satisfies the Lipschitz condition

$$\|f''(x) - f''(y)\| \leq L\,\|x - y\| \quad \forall\, x,\, y \in \mathbb{R}^n,$$

then, for any initial guess, the method converges quadratically. (Try to prove this!)

Exercises

1. Prove that the steepest descent method in the general case can not converge faster than by geometric progression.

2. Suppose f fulfills the conditions of Lemma 18.5.1. Prove that the steepest descent method converges to the minimum point by geometric progression.

   Is it true that

   $$\varepsilon_{k+1} \leq \left(\frac{M - m}{M + m}\right)^2 \varepsilon_k\,?$$

3. A reserve of elementary operations consists of the four arithmetic operations (addition, subtraction, multiplication, and division). Prove that if a functional $f\,(x)$ at x is defined by a simple computation of complexity m, then $f\,(x)$ and $\operatorname{grad} f\,(x)$ at x can be found through a simple computation of complexity not higher than $5\,m$.

4. A reserve of elementary operations comprises additions, subtractions, and multiplications. Prove that if a functional $f(x)$ at x is defined by a simple computation of complexity m, then $f(x)$ and grad $f(x)$ at x can be found through a simple computation of complexity not higher than $3\,m$.

5. A linear computation of complexity m gives the components of the vector $y = A\,x$, $A \in \mathbb{R}^{k \times n}$. Is it true that the components of the vector $z = A^T y$ can be found through a simple computation of the same complexity m ?

Lecture 19

19.1 Quadratic functionals and linear systems

If $f(x) = \frac{1}{2}(Ax, x) - \operatorname{Re}(b, x)$, $A = A^* \in \mathbb{C}^{n \times n}$, then the boundedness of f from below is equivalent to the nonnegative definiteness of A (prove this). Let us assume that $A > 0$. In this case, a linear system $Ax = b$ has a unique solution z, and, for any x,

$$f(x) - f(z) = \frac{1}{2}(A(x - z), x - z) \equiv E(x). \quad \Rightarrow$$

z is the single minimum point for $f(x)$. $\Rightarrow$ A minimization method for f can equally serve as a method of solving a linear system with the Hermitian positively definite coefficient matrix.

The functional $E(x)$ is often referred to as the *error functional* for $Ax = b$. It differs from f only by a constant. So, when minimizing f, we do the same with E.

If A is an arbitrary nonsingular matrix (not necessarily Hermitian), then the solution to $Ax = b$ can be obtained by minimizing the quadratic functional $r(x) = \|b - Ax\|_2^2$. It is called the *residual functional* for $Ax = b$.

19.2 Minimization over the subspace and projection methods

As in the steepest descent method, in many others every iteration requires solving a local minimization problem. The local minimization can be conducted over subspaces of a fixed dimension (equal to 1 in the steepest descent method). But the wider the subspace, the nearer we can get to the desired solution.

In $\mathbb{C}^n$, consider a chain of subspaces

$$L_1 \subset L_2 \subset \ldots \subset L_k \subset \mathbb{C}^n, \quad \dim L_i = i, \quad i = 1, \ldots, k,$$

and for every k go in for $x_k = \operatorname{argmin}\{f(x) : x \in L_k\}$. For a functional f bounded from below, we come to the solution no later than n steps. (Why?)

This idea evolves into a variety of *projection methods*. Suppose it is required to solve $Ax = b$. Then we bring in projectors Q_k, P_k of rank k and assume that the *projection equation*

$$(Q_k\, A\, P_k)\, x \;=\; Q_k\, b, \quad x \;=\; P_k\, x,$$

has a unique solution x_k. We would anticipate that x_k approximates x.

The projection equation is often an equivalent formulation for f to be minimized on the subspace $L_k = \operatorname{im} L_k$.

19.3   Krylov subspaces

Subspaces of the form

$$\mathcal{K}_i \;\equiv\; \mathcal{K}_i\,(b,\,A) \;\equiv\; \operatorname{span}\{b,\,A\,b,\,\ldots,\,A^{i-1}\,b\}, \quad i = 1,\,2,\,\ldots\,,$$

are called the *Krylov subspaces*.

If we intend to solve a system $Ax = b$ then the idea of minimization over the subspace can acquire the following form:

$$x_i \;=\; \operatorname{argmin}\{\|b - A\,x\|_2 : x \in \mathcal{K}_i\}, \quad i = 1,\,2,\,\ldots\,. \qquad (19.3.1)$$

In case the matrix A is nonsingular, the vectors x_i are determined uniquely (prove this). A method of generating the sequence of x_i is referred to as the *method of minimal residuals*.

Why do we know that the method of minimal residuals always leads to the solution of $Ax = b$? Sooner or later, we get $\mathcal{K}_i = \mathcal{K}_{i+1} \Rightarrow A\mathcal{K}_i \subset \mathcal{K}_i$. The matrix A is nonsingular $\Rightarrow A\mathcal{K}_i = \mathcal{K}_i \Rightarrow$ since $b \in \mathcal{K}_i$, for some $x \in \mathcal{K}_i$, $Ax = b$ must hold.

We will discuss some implementation details a little later.

19.4   Optimal subspaces

Suppose that a system $Ax = b$ with a nonsingular matrix A is solved by residual minimization over the subspace. How do we choose the subspaces?

Consider an arbitrary algorithm Φ that generates the subspaces L_i:

$$L_{i+1} \;=\; \Phi\,(b,\,L_i,\,A\,L_i).$$

If $L_i = \operatorname{span}\{p_1,\,\ldots,\,p_i\}$, then a new subspace is defined by a vector

$$p_{i+1} \;=\; \phi_{i+1}\,(b,\,p_1,\,\ldots,\,p_i,\,A\,p_1,\,\ldots,\,A\,p_i).$$

It means that, for every new subspace to appear, we are to perform exactly one matrix-vector multiplication (A by p_i). The same operation's result can be used to minimize the residual over L_i. Therefore, we define the cost for the algorithm Φ as follows:

$$m\,(\Phi,\,A,\,b) \;\equiv\; \min\{i: \min_{y\in L_i} \|b - A\,y\|_2 \;\leq\; \varepsilon\}.$$

A bad algorithm Φ is not prohibited to have $m\,(\Phi,\,A,\,b) = +\infty$.

For an individual matrix and right-hand side, the best would be the algorithm of the minimal cost. However, one and the same algorithm of subspace generation is usually applied to different matrices, and, apparently, we should judge it by the behavior on all the matrices of interest. By the cost of an algorithm Φ on a class of matrices $\mathcal{A}$, we understand its cost in the worst case:

$$m\,(\Phi,\,b) \;\equiv\; \sup_{A\in\mathcal{A}} m\,(\Phi,\,A,\,,b).$$

An algorithm $\hat{\Phi}$ is said to be *optimal* on the class $\mathcal{A}$ if, for any right-hand side b, it is valid that

$$m\,(\hat{\Phi},\,b) \;\leq\; m\,(\Phi,\,b) \quad \forall\ \Phi.$$

In the late 70s, Nemirovski and Yudin (Russia) discovered that the information about a linear system captured by the Krylov subspaces is almost optimal whichever way we use it. We will use it in the following way: having a subspace (in other words, a basis of it), we minimize the residual.

19.5 Optimality of the Krylov subspace

Denote by $\mathcal{K}$ an algorithm generating the Krylov subspaces. It is almost optimal on an arbitrary *unitarily invariant* class of matrices $\mathcal{A}$ (if $A \in \mathcal{A}$, then $Q^*AQ \in \mathcal{A}$ for any unitary matrix Q) in the sense that, for any b,

$$m\,(\mathcal{K},\,b) \;\leq\; 2\,m\,(\Phi,\,b) + 1 \quad \forall\ \Phi. \tag{19.5.1}$$

This follows immediately from the following theorem.

Theorem 19.5.1 *For any nonsingular matrix A and any algorithm Φ of subspace generation, there exists a unitary matrix Q such that*

$$m\,(\mathcal{K},\,A,\,b) \;\leq\; 2\,m\,(\Phi,\,Q^*AQ,b) + 1. \tag{19.5.2}$$

Proof. Following A. Chou[1], we construct a unitary matrix Q providing for $S = Q^*AQ$ to fulfill either $m \equiv m\,(\Phi,\,S,b) = +\infty$ or

$$m\,(\mathcal{K},\,A,\,b) \;\leq\; \dim \operatorname{span}\{b,\,L_m,\,SL_m\}, \tag{19.5.3}$$

---

[1] A.W.Chou. On the optimality of Krylov information. *J. of Complexity* 3: 26–40 (1987).

where L_m is a subspace generated by the algorithm Φ for S.

Consider the orthonormal basis $v_1, \ldots, v_n$ providing that

$$\text{span}\{v_1, \ldots, v_i\} = \mathcal{K}_i(A, b) \ \forall \ i \ : \ \dim \mathcal{K}_i = i.$$

Using the algorithm Φ, we build a sequence of orthonormal vectors $u_1, u_2, \ldots$ and unitary matrices Q_i being arbitrary in all except the conditions

$$Q_i u_j = v_j, \quad 1 \leq j \leq r'(i). \tag{19.5.4}$$

Set $L_0 = \mathcal{M}_0 = \text{span}\{b\} = \text{span}\{u_1\}$, $Q_0 = I$. Assume that we already have some matrices $S_j = Q_j^* A Q_j$, $1 \leq j \leq i$ such that the algorithm Φ generates for S_j the subspaces $L_j = \text{span}\{p_1, \ldots, p_j\}$ (it is important that exactly the same subspaces are obtained for S_i). Consider the following subspaces:

$$\begin{aligned} \mathcal{M}_i &\equiv \text{span}\{b, L_i, S_i L_i\} = \text{span}\{u_1, \ldots, u_{r(i)}\}, \\ \mathcal{M}'_i &\equiv \text{span}\{\mathcal{M}_i, p_{i+1}\} = \text{span}\{u_1, \ldots, u_{r'(i+1)}\}. \end{aligned}$$

Here, p_{i+1} is generated by the algorithm Φ for S_i. Let the unitary matrix Q_{i+1} enjoy the conditions (19.5.4). Then $S_{i+1} = Q_{i+1}^* A Q_{i+1}$ and, by definition,

$$\mathcal{M}_{i+1} = \text{span}\{\mathcal{M}_i, p_{i+1}, S_{i+1} p_{i+1}\} = \text{span}\{u_1, \ldots, u_{r(i+1)}\}.$$

If $r(i+1) > r(i)$, then $r'(i+1) = r(i) + 1$. In this case, we require that $u_{r(i+1)} = Q_{i+1}^* v_{r(i+1)}$. The vector defined in such a way is orthogonal to every u_j for $j < r(i+1)$. At the same time,

$$u_{r(i+1)} = Q_{i+1}^* v_{r(i+1)} \in Q_{i+1}^* A Q_{i+1} \mathcal{M}'_{i+1} \subset \mathcal{M}_{i+1}.$$

Now, let us assume that $\exists \ y \in L_m \ : \ \|b - S_m y\|_2 \leq \varepsilon$. Then

$$\|b - S_m y\|_2 \geq \min_{u \in \mathcal{M}'_{m-1}} \|b - A Q_m u\| = \min_{v \in \mathcal{K}_{r'(m)}} \|b - A v\|,$$

and now, (19.5.3) follows from $r'(m) \leq \dim \mathcal{M}_m$.

If no m provides the accuracy ε for S_m on the mth step, then sooner or later, $Q_m = Q_{m+1} = \ldots$. Consequently, for this S_m, the accuracy ε is never reached, no matter how long we iterate, and hence, $m(\Phi, S_m, b) = +\infty$. $\square$

19.6   Method of minimal residuals

To solve a system $Ax = b$ with a nonsingular matrix A, we choose an (arbitrary) initial guess x_0 and, then, proceed, in fact, to a reduced system $Au = r_0$, where $r_0 = b - Ax_0$, $x = x_0 + u$. In the method, we successively minimize the residual on the Krylov subspaces for the reduced system (provided that $r_0 \neq 0$).

There are several ways to implement this.[2] I think it is not a big deal to find one of your own, is it?

Anyway, we go on with those who still want more details. Let $q_1 = r_0/||r_0||_2$. To have an orthonormal basis $q_1, \ldots, q_{i+1}$ in $\mathcal{K}_{i+1} = \mathcal{K}_{i+1}(r_0, A)$, we orthogonalize the vector $A q_i$ to the preceding vectors $q_1, \ldots, q_i$. If $Q_i \equiv [q_1 \ldots q_i] \in \mathbb{C}^{n \times i}$, then we obtain

$$A Q_i = Q_{i+1} \hat{H}_i, \quad \hat{H}_i = \left[\begin{array}{c} H_i \\ 0 \ldots 0 \; h_{i+1\,i} \end{array} \right],$$

where H_i is an upper Hessenberg matrix of order i.

Further, consider the QR decomposition for the rectangular matrix $\hat{H}_i$:

$$\hat{H}_i = U_i R_i, \quad R_i \in \mathbb{C}^{i \times i}.$$

Then the residual $||r_0 - AQ_i y||_2$ takes its minimum over all y if $y = y_i$ satisfies the equation (prove this)

$$R_i y_i = z_i \equiv ||r_0||_2 U_i^* e_1, \quad \text{where} \quad e_1 = [1 \; 0 \ldots 0]^T.$$

The matrix R_i is nonsingular. (Why?) Therefore,

$$x_i = x_0 + Q_i y_i = x_0 + Q_i R_i^{-1} z_i.$$

Note that the Hessenberg matrices H_i are the leading principal submatrices of the largest Hessenberg matrix coupled with the last iteration. Apart from multiplying a matrix by a vector, other overheads of the ith iteration are mostly due to orthogonalization. Beside it, we still have some work for computing the vectors z_i and then y_i. Thanks to the Hessenberg structure of the matrix H_i, it takes only $\mathcal{O}(i^2)$ operations.

More details can be found in the rather recent, well-cited paper of Y. Saad and M. N. Schultz.[3] It is probably in this work that the name "GMRES" was given. By a "nongeneralized yet" method of minimal residuals, one usually means the method of minimal residuals in the case of a Hermitian matrix. If so, the matrices H_i turn out to be tridiagonal. (Prove this!)

For a large number of iterations, the arithmetic costs and memory required to retain the vectors q_i can get forbiddingly high. For such cases, one applies the method of minimal residuals *with restarts*: in advance, we set the maximal allowed dimension of the Krylov subspace, and iterate until the stopping criterion bids us to quit or the maximal subspace is reached. In the latter case, we take the approximate solution at hand as a new initial guess and start to generate a new chain of Krylov subspaces.

---

[2] G. I. Marchuk and Yu. A. Kuznetsov. Iterative methods and quadratic functional. *Methods of Numerical Mathematics*. Novosibirsk, 1975, pp. 4–143.

[3] Y. Saad and M. H. Schultz. GMRES: A generalized minimal residual algorithm for solving nonsymmetric linear systems. *SIAM J. Scientific and Stat. Comp.* 7: 856–869 (1986).

19.7 A-norm and A-orthogonality

Let A be a Hermitian positive definite matrix of order n. Then, for any pair of vectors x, $y \in \mathbb{C}^n$, let us set

$$(x, y)_A \equiv (Ax, y).$$

This is a scalar product in $\mathbb{C}^n$ (check this).

The norm $||x||_A \equiv \sqrt{(x, y)_A}$ is called the A-norm. Vectors x are y said to be A-orthogonal if $(x, y)_A = 0$. Any two orthogonal vectors enjoy Pythagor's theorem: $||x + y||_A^2 = ||x||_A^2 + ||y||_A^2$.

19.8   Method of conjugate gradients

Suppose z is the solution to a system $Ax = b$ with $A = A^* > 0$. Choose a guess x_0, take $r_0 = b - Ax_0$ to build up $\mathcal{K}_i = \mathcal{K}_i(r_0, A)$, and let $x_i = x_0 + y_i$, where $y_i \in \mathcal{K}_i$. The only difference with the method of minimal residuals is that now we minimize over $\mathcal{K}_i$ the A-norm of the error $e = x - z$:

$$y_i = \operatorname{argmin} \{||x_0 + y - z||_A : y \in \mathcal{K}_i\}.$$

By Pythagor's theorem,

$$(x_i, y)_A = (z, y)_A \ \forall \ y \in \mathcal{K}_i \quad \Leftrightarrow \quad r_i \equiv b - A x_i \perp \mathcal{K}_i.$$

It is best if we have an A-orthogonal basis $p_1, \dots, p_i$ in $\mathcal{K}_i$. In this case, the expansion

$$y_i = \alpha_{1i} p_1 + \dots + \alpha_{ii} p_i$$

has coefficients independent of the second index i, i.e., $\alpha_{ji} \equiv \alpha_j$. Therefore (do not forget that $r_i \perp \mathcal{K}_i$),

$$x_i = x_{i-1} + \alpha_i p_i \ \Rightarrow \ r_i = r_{i-1} - \alpha_i A p_i \ \Rightarrow \ \alpha_i = \frac{(r_{i-1}, p_i)}{(A p_i, p_i)}.$$

If $r_i \neq 0$, then write $p_{i+1} = r_i + \beta_{i1} p_1 + \dots + \beta_{ii} p_i$. Next, $r_i \perp \mathcal{K}_i \Rightarrow \beta_{ij} = 0$ for $j < i$. (Prove this!) $\Rightarrow$

$$p_{i+1} = r_i + \beta_i p_i, \quad \beta_i = \frac{(r_i, A p_i)}{(A p_i, p_i)}.$$

Since $p_i = r_{i-1} + \beta_{i-1} p_{i-1}$, we find that $\alpha_i = \frac{(r_{i-1}, r_{i-1})}{(A p_i, p_i)}$. If $r_{i-1} \neq 0$, then $\alpha_i \neq 0$. $\Rightarrow$

$$(r_i, A p_i) = (r_i, \frac{r_{i-1} - r_i}{\alpha_i}) = -\frac{(r_i, r_i)}{\alpha_i} \ \Rightarrow \ \beta_i = \frac{(r_i, r_i)}{(r_{i-1}, r_{i-1})}.$$

Finally, we obtain the following formulas of the *method of conjugate gradients*:

$$\begin{aligned}
\alpha_i &= (r_{i-1}, r_{i-1})/(Ap_i, p_i), \\
x_i &= x_{i-1} + \alpha_i\, p_i, \\
r_i &= r_{i-1} - \alpha_i\, Ap_i, \\
\beta_i &= (r_i, r_i)/(r_{i-1}, r_{i-1}), \\
p_{i+1} &= r_i + \beta_i\, p_i.
\end{aligned} \qquad (19.8.1)$$

A wonderful thing is that the recurrence relations now are *short*: when minimizing over the subspace $\mathcal{K}_i$, we do it without a full basis!

19.9 Arnoldi method and Lanczos method

Given a vector $r_0 \neq 0$, set i as the minimal index such that $\mathcal{K}_i(r_0, A) = \mathcal{K}_{i+1}(r_0, A)$. In the subspace $\mathcal{K}_i$, we consider an orthonormal basis $q_1, \ldots, q_i$ providing that span $\{q_1, \ldots, q_j\} = \mathcal{K}_j$ for all $1 \leq j \leq i$. The vector q_{j+1} can be built through orthogonalization of the vector $A\, q_j$ to the preceding vectors $q_1, \ldots, q_j$.

Set $Q_j = [q_1\ \ldots\ q_j]$. Then the matrix

$$A_j = Q_j^* A Q_j$$

is called the *projective restriction* of A on $\mathcal{K}_j$. For $1 \leq j \leq i$, the matrix A_j is a leading principal submatrix of A_i.

In the general case, A_i is an upper Hessenberg matrix. If A is Hermitian, then A_i is a Hermitian tridiagonal matrix. If a method constructs the matrices A_j and Q_j, it is associated with the name of Arnoldi, in the general case, and with Lanczos in the Hermitian case.

Under the assumptions made, $\mathcal{K}_i$ is A-invariant. Hence, $\lambda(A_i) \subset \lambda(A)$. However, the eigenvalues of the projective restrictions A_j can approximate the eigenvalues of A well enough with j being sometimes far less than i (we explain this in the following lecture). Such an approach to computing eigenvalues is attributed to Arnoldi, in the general case, and to Lanczos in the Hermitian case.

19.10 Arnoldi and Lanczos without Krylov

In the methods of Arnoldi and Lanczos, Krylov subspaces are referenced to through the orthonormal bases constructed for them. But curiously enough, one can glean the same bases by analyzing some well-known matrix decompositions without giving a thought to Krylov's subspaces.

We know that a matrix $A \in \mathbb{C}^{n \times n}$ is unitarily similar to some upper Hessenberg matrix H:

$$A\,[q_1\ \ldots\ q_n] = [q_1\ \ldots\ q_n]\, H.$$

Choose q_1 to be arbitrary in all except for the condition $||q_1||_2 = 1$, and equate the first columns:

$$A\,q_1 \;=\; q_1\,h_{11} + q_2\,h_{21}.$$

Since q_1 and q_2 are orthogonal, $h_{11} = (Aq_1, q_1)$. Then the vector q_2 is obtained from $A\,q_1 - q_1\,h_{11}$ by normalization (provided that the vector is not zero). Next, we equate the second columns and find q_3, and so on.

We stop at the ith column, if the candidate for q_{i+1} was zero. In this case, however,

$$A\,[q_1 \;\ldots\; q_i] \;=\; [q_1 \;\ldots\; q_i]\,H_i,$$

where H_i is the leading $i \times i$ submatrix in H. Note that the subspace $\operatorname{im} H_i$ is A-invariant.

Thus, we have the subspaces

$$L_j \;=\; \operatorname{span}\{q_1, \ldots, q_j\}$$

and can think of using them in the projection methods.

Formally, we did without the Krylov subspace definition, but not without the subspaces themselves. If we still want to remember that definition, we should confess that $L_j \;=\; \mathcal{K}_j\,(q_1, A)$.

19.11   From matrix factorizations to iterative methods

The method of conjugate gradients can be derived directly from the Choleski decomposition.[4]

Let $A = A^* > 0$. Then $A_j = Q_j^* A\,Q_j$ is a Hermitian tridiagonal matrix (we use the notation from the section 19.9). Moreover, $A_j > 0$ (why?) and we may consider its Choleski decomposition $A_j = R_j^* R_j$. Since A_j is tridiagonal, the upper triangular matrix R_j is bidiagonal:

$$R_j \;=\; \begin{bmatrix} \gamma_1 & \delta_1 & & & & \\ & \gamma_2 & \delta_2 & & & \\ & & \ddots & \ddots & & \\ & & & \ddots & \ddots & \\ & & & & \gamma_{j-1} & \delta_{j-1} \\ & & & & & \delta_j \end{bmatrix}.$$

Take up a nonsingular diagonal (beside this, arbitrary) matrix D_j and set

$$P_j \;=\; [p_1 \;\ldots\; p_j] \;=\; Q_j\,R_j^{-1}\,D_j \quad \Rightarrow \quad P_j^*\,A\,P_j \;=\; D_j^*\,D_j. \quad \Rightarrow$$

---

[4]I first heard about this striking relation from S. A. Goreinov.

The columns of the matrix P_j are A-orthogonal, and we can see that

$$p_{j+1} \frac{\gamma_{j+1}}{d_{j+1}} + p_j \frac{\delta_j}{d_j} = q_{j+1} \quad \Rightarrow \quad p_{j+1} = \frac{d_{j+1}}{\gamma_{j+1}} q_{j+1} + \frac{-\delta_j}{\gamma_{j+1}} \frac{d_{j+1}}{d_j} p_j.$$

Meanwhile, we want to solve a system $Ax = b$. Let $x_j = x_0 + P_j y$ and $r_j \perp \operatorname{im} P_j$ (the latter is equivalent to minimizing the error in the A-norm). This implies that $x_j = x_{j-1} + \alpha_j p_j$, $r_j = r_{j-1} - \alpha_j A p_j$. What is more, if $r_j \neq 0$, then q_{j+1} is collinear with r_j. Therefore, if we build the vectors $q_1, \dots, q_{j+1}$ starting from $q_1 = r_0 / \|r_0\|_2$, then nothing prevents us from choosing $q_{j+1} = r_j / \|r_j\|_2$ and then specifying D_j in the following way:

$$d_1 = \|r_0\|_2; \quad d_{j+1} = \gamma_{j+1} / \|r_j\|_2, \ j = 0, 1, \dots \ .$$

All in all, the expression for p_{j+1} acquires the familiar form

$$p_{j+1} = r_j + \beta_j p_j.$$

Other matrix factorizations can lead to other useful iterative algorithms. For example, the unitary bidiagonalization of an arbitrary matrix transforms into the Lanczos bidiagonalization method that allows one to compute the singular values and vectors. Every iteration of this method will require two matrix-vector multiplications, one with A, and the other with A^*.

19.12 Surrogate scalar product

A nice thing is that the conjugate gradient method (and the minimal residual one) for Hermitian matrices can be performed through *short recurrences*. With a certain price paid, however, we can also rejuvenate short recurrences for non-Hermitian matrices.

To cover more applications, we sometimes replace the scalar product by more convenient bilinear or sesquilinear forms. Take a matrix $D \in \mathbb{C}^{n \times n}$ and introduce a *surrogate scalar product* in one of the following ways:

$$\langle x, y \rangle = y^T D x, \quad x, y \in \mathbb{C}^n, \tag{*}$$

or, alternatively,

$$\langle x, y \rangle = y^* D x, \quad x, y \in \mathbb{C}^n. \tag{**}$$

For a matrix $A \in \mathbb{C}^{n \times n}$, denote by A' a *dual matrix* such that $\langle Ax, y \rangle = \langle x, A'y \rangle \ \forall \ x, y \in \mathbb{C}^n$. If α is a scalar quantity, then α' designates a dual one providing that $\langle \alpha x, y \rangle = \langle x, \alpha' y \rangle \ \forall \ x, y \in \mathbb{C}^n$. From now on, we write $x \perp y$ if $\langle x, y \rangle = 0$.

For the time being, set $D = I$. Then, $A' = A^T$ for the case (*), and $A' = A^*$ for (**).

19.13   Biorthogonalization approach

To solve $Ax = b$ with a nonsingular $A \in \mathbb{C}^{n \times n}$, we start from some x_0 and try to approximate x by a vector $x_i = x_0 + y_i$, where $y_i \in \mathcal{K}_i(r_0, A) = $ span $\{p_1, \ldots, p_i\}$, $r_0 = b - Ax_0$, postulating that $r_i \equiv b - Ax_i \perp \mathcal{K}_i(r_0', A') = $ span $\{p_1', \ldots, p_i'\}$, where r_0' is chosen at the beginning so that $\langle r_0, r_0' \rangle \neq 0$.

Assume that $p_1, \ldots, p_i$ and $p_1', \ldots, p_i'$ are A-biorthogonal in the sence that $\langle Ap_j, p_k' \rangle = 0$ for all $j \neq k$. Let it not vanish for $j = k$. Then from
$$r_i = r_0 - Ax = r_0 - \sum_{j=1}^{i} \alpha_{ji} Ap_j \text{ it follows that } \alpha_{ji} = \alpha_j = \langle r_0, p_j' \rangle / \langle Ap_j, p_j \rangle.$$
$\Rightarrow x_i = x_{i-1} + \alpha_i p_i \Rightarrow r_i = r_{i-1} - \alpha_i Ap_i$. If $r_{i-1} \perp \mathcal{K}_{i-1}(r_0', A')$ (we assume this), then $\alpha_i = \langle r_{i-1}, p_i' \rangle / \langle Ap_i, p_i' \rangle$.

Consider also the vectors $r_j' = r_{j-1}' - \tilde{\alpha}_j' A' p_j'$, stipulating that they satisfy $\mathcal{K}_j(r_0, A) \perp r_j'$ for all $j \leq i$. If the latter already holds for all $j < i$, then, we have to take $\tilde{\alpha}_i = \langle p_i, r_{i-1}' \rangle / \langle Ap_i, p_i' \rangle$.

Assume that $\alpha_j, \tilde{\alpha}_j \neq 0$ for all $j \leq i$. Then, we can take for granted that

$$p_{i+1} = r_i + \beta_i p_i, \qquad p_{i+1}' = r_i' + \tilde{\beta}_i' p_i'.$$

With that much, we immediately conclude that $\alpha = \tilde{\alpha} = \langle r_{i-1}, r_{i-1}' \rangle / \langle Ap_i, p_i' \rangle$. Since $\langle Ar_i, p_j' \rangle = \langle r_i, A'p_j' \rangle = 0$ for $j < i$, we maintain the biorthogonality property by choosing $\beta_i = - \langle Ar_i, p_i' \rangle / \langle Ap_i, p_i' \rangle$. Next,

$$\langle Ar_i, p_i' \rangle = \langle r_i, A'p_i' \rangle = \langle r_i, \frac{r_{i-1}' - r_i'}{\alpha_i} \rangle = \frac{-\langle r_i, r_i' \rangle}{\alpha_i} \quad \Rightarrow \quad \beta_i = \frac{\langle r_i, r_i' \rangle}{\langle r_{i-1}, r_{i-1}' \rangle}.$$

It is easy to verify that $\tilde{\beta}_i = \beta_i$.

Finally, we arrive at the so-called *biorthogonal conjugate gradient algorithm*:

$$r_0 = b - Ax_0, \quad p_1 = r_0, \quad \text{choose } r_0', \quad p_1' = r_0';$$
$$\alpha_i = \langle r_{i-1}, r_{i-1}' \rangle / \langle Ap_i, p_i' \rangle,$$
$$x_i = x_{i-1} + \alpha_i p_i, \quad r_i = r_{i-1} - \alpha_1 Ap_i, \quad r_i' = r_{i-1}' - \alpha_1' A'p_i',$$
$$\beta_i = \langle r_i, r_i' \rangle / \langle r_{i-1}, r_{i-1}' \rangle,$$
$$p_{i+1} = r_i + \beta_i p_i, \quad p_{i+1}' = r_i' + \beta_i' p_i', \quad i = 1, 2, \ldots.$$

Note that $\alpha_i' = \alpha_i$ and $\beta_i' = \beta_i$ for the surrogate scalar product of the type $(*)$, and we turn to the complex conjugate values for the case $(**)$.

19.14   Breakdowns

In the biorthogonal conjugate gradient algorithm, one foresees the possibility of *breakdowns* of the following two types:

- $\langle Ap_i, p_i' \rangle = 0$ for some i;

- $\langle r_{i-1}, r_{i-1}' \rangle = 0 \implies \alpha_i = 0$ whereas $r_i \neq 0$.

In fighting against the breakdowns, what can help really are some *block versions* of the above algorithm.

19.15 Quasi-Minimization idea

For short recurrences in the non-Hermitian case, the price we pay is the loss of the minimization property. It results in a teasing chaotic behavior of the residual values, which makes it difficult to decide whether to quit or not. However, we can move from the vectors x_i to some other, more agreeable vectors $\hat{x}_i$ of which the residuals $\hat{r}_i$ behave quite decently.

Let $P_i = [p_1, \ldots, p_i]$, $\rho_i = \|r_i\|_2$, and $R_i = [r_0/\rho_0, \ldots, r_i/\rho_i]$. If we write $\hat{x}_i = x_0 + P_i y$, then

$$\hat{r}_i = r_0 - AP_i y = R_i v_i$$

with the vector v_i of the form

$$v_i = [\rho_0(1 - \xi_0), \ \rho_1(\xi_1 - \xi_0), \ \ldots, \ \rho_{i-1}(\xi_i - \xi_{i-1}), \ \rho_i \xi_i]^T$$

for some $\xi_0, \ldots, \xi_i$. Let us choose these quantities to provide the least possible value for $\|v\|_2$. Since $\|R_i\|_2 \leq \sqrt{i+1}$ (why?),

$$\|\hat{r}_i\|_2 \leq \sqrt{i+1} \min \|v_i\|_2.$$

This is what we call the *quasi-minimization property*.

Set $\eta_0 = 1 - \xi_0$, $\eta_1 = \xi_1 - \xi_0$, $\ldots$, $\eta_{i-1} = \xi_i - \xi_{i-1}$, $\eta_i = \xi_i$. Then, we are to minimize the functional $f = \|v\|_2^2 = \rho^2 \eta_0^2 + \ldots \rho_i^2 \eta_i^2$ subject to the constraint $\eta_0 + \ldots + \eta_i = 1$. By the standard technique using Lagrange's multipliers,[5] we find that

$$\eta_j = \frac{\sigma_j}{s_i},$$

where

$$\sigma_j = \frac{1}{\rho_j^2}, \qquad s_i = \sum_{j=1}^{i} \frac{1}{\rho_j^2}. \tag{19.15.1}$$

Consequently,

$$\hat{x}_i = \sum_{j=0}^{i} \frac{\sigma_j}{s_i} x_i, \qquad \hat{r}_i = \sum_{j=0}^{i} \frac{\sigma_j}{s_i} r_i.$$

---

[5]L. Zhou and H. F. Walker. Residual smoothing techniques for iterative methods. *SIAM J. on Sci. Comput.* 15 (2): 297–312 (1994).

These formulas are not for computation yet. We use them to infer that

$$\hat{x}_i = \frac{s_{i-1}}{s_i}\,\hat{x}_i + \frac{\sigma_i}{s_i}\,x_i, \qquad \hat{r}_i = \frac{s_{i-1}}{s_i}\,\hat{r}_i + \frac{\sigma_i}{s_i}\,r_i. \tag{19.15.2}$$

We are much better off, indeed, by supplementing the biorthogonal conjugate gradient algorithm with (19.15.1) and (19.15.2).

The quasi-minimization idea can also be useful for some other iterative algorithms. It has been widely discussed since the recent work of R. W. Freund and N. M. Nachtigal.[6]

Exercises

1. In theory, the residual in the method of minimal residuals goes down monotonically. In practice, it does until a restart, when one might observe it increasing. Do you have an idea why this happens?

2. What vectors can be simultaneously orthogonal and A-orthogonal?

3. Prove that, in the method of conjugate gradients, the residual of a step i is A-orthogonal to the residual of any step j so long as $|j - i| > 1$.

4. Prove that, in the biorthogonal conjugate gradient method, the vectors r_i and r'_j are biorthogonal in the sence that $\langle r_i, r'_j \rangle = 0$ for $i \neq j$.

---

[6]R. W. Freund and N. M. Nachtigal. QMR: a quasi-minimal residual method for non-Hermitian linear systems. *Numer. Math.* 60: 315–339 (1991).

Lecture 20

20.1 Convergence rate of the conjugate gradient method

Consider the expansion of the initial residual $r_0 = \sum_{i=1}^{k} \xi_i z_i$ in an orthonormal eigenvector system of the given $n \times n$ matrix $A = A^* > 0$.

Then, the method of conjugate gradients arrives at the solution to $Ax = b$ no later than the kth step. If $A z_i = \lambda_i z_i$ and $\lambda_i \neq \lambda_j$ for $i \neq j$, $1 \leq i, j \leq k$, then the solution emerges exactly on the kth step. (Prove this.)

It is remarkable that frequently the approximate solution becomes accurate enough pretty much before the kth step.

Note that $p_i = \phi_{i-1}(A) r_0$, where $\phi_{i-1}(\lambda)$ is a polynomial of degree $i - 1$. (Why?) It implies that $r_i = \psi_i(A) r_0$, where $\psi_i(\lambda) = 1 + \lambda \phi_{i-1}(\lambda)$ ($\Rightarrow$ $\psi_i(0) = 1$). Since the conjugate gradient method on the ith step minimizes the A-norm of the error $e_i = x_i - x$ (x is the exact solution), we find that

$$
\begin{aligned}
\|e_i\|_A^2 &= (Ae_i, e_i) = (r_i, A^{-1} r_i) \\
&= \sum_{j=1}^{k} \frac{\psi_i^2(\lambda_j)}{\lambda_j} \xi_j^2 \leq \left(\max_{1 \leq j \leq k} |\psi_i(\lambda_j)| \right)^2 \|e_0\|_A^2.
\end{aligned}
$$

If $\lambda(A) \subset [m, M]$ then, obviously, for any polynomial $\psi_i(\lambda)$ of degree not higher then i with constant term equal to 1, the error can be estimated from the above as follows:

$$
\|e_i\|_A \leq \max_{m \leq \lambda \leq M} |\psi_i(\lambda)| \|e_0\|_A. \tag{20.1.1}
$$

20.2   Chebyshev polynomials again

Let $T_i(t)$ be the Chebyshev polynomial of degree i. Take into account that

$$t \in [-1, 1] \quad \Leftrightarrow \quad \lambda = \frac{m+M}{2} + \frac{M-m}{2} t \in [m, M],$$

and consider a polynomial

$$T_{i;\, m,\, M}(\lambda) = T_i \left(\frac{2\lambda - m - M}{M - m} \right).$$

We call it the *Chebyshev polynomial for the segment* $[m, M]$.

Obviously, we may take

$$\psi_i(\lambda) = T_{i;\, m,\, M}(\lambda)/T_{i;\, m,\, M}(0).$$

Then, since $T_{i;\, m,\, M}(0) = 1/T_i(-\frac{M+m}{M-m})$, we obtain

$$\max_{m \leq \lambda \leq M} |\psi_i(\lambda)| \leq \frac{1}{|T_i(-\frac{M+m}{M-m})|}.$$

Thus,

$$\|e_i\|_A \leq \frac{1}{|T_i(-\frac{M+m}{M-m})|} \|e_0\|_A. \qquad (20.2.1)$$

This estimate seems attractive, for the Chebyshev polynomials grow exponentially in i so long as $|t| > 1$:

$$T_i(t) = \frac{1}{2} \left(t + \sqrt{t^2 - 1} \right)^i + \frac{1}{2} \left(t - \sqrt{t^2 - 1} \right)^i.$$

Allowing for $|T_i(t)| \geq \frac{1}{2}|t|^i$, we can derive from there, for example, the estimate

$$\|e_i\|_A \leq 2 \left(\frac{M-m}{M+m} \right)^i \|e_0\|_A, \qquad (20.2.2)$$

which, we dare note, cannot satisfy us very much, because it is not any better than the steepest decent method estimate. But still do not make haste to draw conclusions.

20.3   Classical estimate

We need to estimate $T_i(t)$ at $t = -(1+\nu)/(1-\nu)$, where $\nu = m/M$.

$$\Rightarrow \quad t^2 - 1 = \frac{(1+\nu)^2 - (1-\nu)^2}{(1-\nu)^2} = \frac{4\nu}{(1-\nu)^2} \quad \Rightarrow$$

$$\sigma \equiv |t| + \sqrt{t^2 - 1} = \frac{1 + \nu + 2\sqrt{\nu}}{1 - \nu} = \frac{(1 + \sqrt{\nu})^2}{1 - \nu} = \frac{1 + \sqrt{\nu}}{1 - \sqrt{\nu}}.$$

Allowing for

$$2\,|T_i(t)| \geq \frac{1 + \sigma^{2i}}{\sigma^i},$$

we find that

$$\|e_i\|_A \leq 2 \left(\frac{1 - \sqrt{\frac{m}{M}}}{1 + \sqrt{\frac{m}{M}}} \right)^i \|e_0\|_A. \tag{20.3.1}$$

The estimate (20.3.1) is better in a big way than (20.2.2). It confirms quantitatively what has been felt quite clear: the method of conjugate gradients is bound to converge *much faster* than the steepest descent method.

20.4 Tighter estimates

The estimate (20.3.1) suggests that the convergence rate slows down for ill-conditioned matrices. However, it could be fast enough even for such cases.[1] Let us try to explain why.

Denote by $\lambda_1 \geq \ldots \geq \lambda_n$ the eigenvalues of $A = A^* > 0$.

Assume that $\lambda_1 \gg \lambda_2$. In this case, the conjugate gradient method behaves as though the condition number of the matrix is equal to λ_2/λ_n. Indeed, we use (20.1.1) with the polynomial

$$\psi_i(\lambda) = \frac{T_{i-1}\left(\frac{2\lambda - \lambda_2 - \lambda_n}{\lambda_2 - \lambda_n} \right)}{T_{i-1}\left(-\frac{\lambda_2 + \lambda_n}{\lambda_2 - \lambda_n} \right)} \left(1 - \frac{\lambda}{\lambda_1} \right). \quad \Rightarrow$$

$$\|e_i\|_A \leq 2 \left(1 - \frac{\lambda_n}{\lambda_1} \right) \left(\frac{1 - \sqrt{\frac{\lambda_n}{\lambda_2}}}{1 + \sqrt{\frac{\lambda_n}{\lambda_2}}} \right)^{i-1} \|e_0\|_A. \tag{20.4.1}$$

If $\lambda_n \ll \lambda_{n-1}$, then, from the standpoint of quality, we have the same result: the method ignores the smallest eigenvalue and behaves as though the condition number is equal to λ_1/λ_{n-1}. In this case (prove this),

$$\|e_i\|_A \leq 2 \frac{\lambda_1}{\lambda_n} \left(\frac{1 - \sqrt{\frac{\lambda_{n-1}}{\lambda_1}}}{1 + \sqrt{\frac{\lambda_{n-1}}{\lambda_1}}} \right)^{i-1} \|e_0\|_A. \tag{20.4.2}$$

All in all, when well-separated from others, the extreme eigenvalues do not much affect the method's behavior. One might still foresee that, in machine arithmetic, the ill condition does not render the convergence picture that exquisite.

---

[1] O. Axelsson and G. Lindskog. The rate of convergence of the conjugate gradient method. *Numer. Math.* 48: 499–523 (1986).

20.5   "Superlinear" convergence and "vanishing" eigenvalues

Strictly speaking, the notion of superlinear convergence rate is not applicable to a process with a finite number of iterations. However, the conjugate gradient method possesses some of its features: the ratio $\omega_i \equiv \|e_i\|_A / \|e_{i-1}\|_A$ usually tends to go down (not monotonically), whereas in linear convergence, one should expect that $\omega_i \approx$ const).

In 1986, A. van der Sluis and H. A. van der Vorst[2] discovered that (in exact arithmetic) the method of conjugate gradients commences since some time to behave as though the matrix A keeps no longer its extreme eigenvalues λ_1 and λ_n, then after some time to have passed it starts to behave as though A has lost, to boot, λ_2 and λ_{n-1}, and so on.

An eigenvalue "vanishes" at the very moment when it is approximated well enough by some eigenvalue of the projective restriction matrix that would be generated (virtually) by the Lanczos method starting from the vector $q_1 = r_0 / \|r_0\|_2$.

Prior to enunciating the corresponding theorem, we need more insight into the Lanczos method and how it relates to the conjugate gradients.

20.6   Ritz values and Ritz vectors

The projective restriction of A on $\mathcal{K}_i(r_0, A)$ is defined as follows:

$$A_i = Q_i^* A Q_i, \quad Q_i = [q_1, \ldots, q_i] = [\frac{r_0}{\|r_0\|_2}, \ldots, \frac{r_{i-1}}{\|r_{i-1}\|_2}]$$

(in the conjugate gradient method, $r_j \perp \mathcal{K}_j$, and the iteration halts with the zero residual).

Denote by $\theta_1 \geq \ldots \geq \theta_i$ the eigenvalues of A_i. They are usually called the *Ritz values*. If $A_i v_j = \theta_j v_j$, $\|v_j\|_2 = 1$, then the vector $y_j = Q_i v_j$ is termed the *Ritz vector* coupled with θ_j.

It is easy to see that $0 = Q_i^* (A Q_i v_j - \theta_j Q_j v_j) = Q_i^* (A y_j - \theta_j y_j)$. Consequently, $A y_j - \theta_j y_j \perp \mathcal{K}_i$ for all $1 \leq i, j \leq i$.

---

[2] A. van der Sluis and H.A. van der Vorst. The rate of convergence of conjugate gradients. *Numer. Math.* 48: 543–560 (1986).

20.7 Convergence of Ritz values

To begin with, why does $\theta_1 \approx \lambda_1$? To answer this question, let us write

$$\theta_1 = \max_{\substack{y \in \mathcal{K}_i \\ y \neq 0}} \frac{(Ay, y)}{(y, y)} = \max_{\phi_{i-1}} \frac{(A\phi_{i-1}(A) r_0, \phi_{i-1}(A) r_0)}{(\phi_{i-1}(A) r_0, \phi_{i-1}(A) r_0)},$$

where the maximum is taken over all polynomials ϕ_{i-1} of degree not higher than $i-1$, provided that $\phi_{i-1}(A) r_0 \neq 0$. Obviously, $\theta_1 \leq \lambda_1$. (Why?) Using the initial residual's expansion $r_0 = \sum_{j=1}^{n} \xi_j z_j$ in the orthonormal eigenvectors of A, we find that

$$\lambda_1 - \theta_1 \leq \lambda_1 - \frac{(A\phi_{i-1}(A) r_0, \phi_{i-1}(A) r_0)}{(\phi_{i-1}(A) r_0, \phi_{i-1}(A) r_0)} = \lambda_1 - \frac{\sum_{k=1}^{n} \lambda_k |\phi_{i-1}(\lambda_k)|^2 |\xi_k|^2}{\sum_{k=1}^{n} |\phi_{i-1}(\lambda_k)|^2 |\xi_k|^2}$$

$$= \frac{\sum_{k=2}^{n} (\lambda_1 - \lambda_k) |\phi_{i-1}(\lambda_k)|^2 |\xi_k|^2}{|\phi_{i-1}(\lambda_1)|^2 |\xi_1|^2 + \sum_{k=2}^{n} |\phi_{i-1}(\lambda_k)|^2 |\xi_k|^2} \leq (\lambda_1 - \lambda_n) \frac{\max_{2 \leq k \leq n} |\phi_{i-1}(\lambda_k)|^2}{|\phi_{i-1}(\lambda_1)|^2} \gamma,$$

$$\gamma = \gamma(r_0) = \sum_{k=2}^{n} |\xi_k|^2 / |\xi_1|^2.$$

The inequality obtained remains valid for an arbitrary polynomial of degree ϕ_{i-1} or less. A good idea is to try such a polynomial of which the value at the point λ_1 is much greater than the values at the points $\lambda_2, \ldots, \lambda_n$. To this end, we can take the familiar Chebyshev polynomial for the segment $[\lambda_n, \lambda_2]$. The choice $\psi_{i-1}(\lambda) = T_{i-1;\lambda_n,\lambda_2}(\lambda)$ results in the following Kaniel–Paige estimate (check this):

$$\lambda_1 - \theta_1 \leq \frac{(\lambda_1 - \lambda_n)}{T_{i-1}^2 (1 + 2\mu)} \gamma(r_0), \quad \mu = \frac{\lambda_1 - \lambda_2}{\lambda_2 - \lambda_n}. \tag{20.7.1}$$

20.8 An important property

Lemma 20.8.1 *Suppose $A = A^* \in \mathbb{C}^{n \times n}$ is reduced via a unitary matrix $Q = [q_1 \ldots q_n]$ to a Hermitian tridiagonal matrix*

$$A = Q^* A Q = \begin{bmatrix} \alpha_1 & \beta_1^* & & & & \\ \beta_1 & \alpha_2 & \beta_2^* & & & \\ & \ddots & \ddots & \ddots & & \\ & & \ddots & \ddots & \ddots & \\ & & & \beta_{n-2} & \alpha_{n-1} & \beta_{n-1}^* \\ & & & & \beta_{n-1} & \alpha_n \end{bmatrix}.$$

Assume that $\beta_j \neq 0$ for $1 \leq j \leq i$. Then $q_{i+1} = \pi_i(A)q_1$ for some polynomial π_i of degree i, which coincides up to normalization with the characteristic polynomial of A_i, the leading submatrix of order i in the matrix A.

Proof. Clearly,

$$A[q_1 \ldots q_i] = [q_1 \ldots q_i]A_i + \beta_i q_{i+1} e_i^T, \qquad e_i^T = [0 \ldots 0\, 1],$$

and, with this, we can choose the polynomials $\pi_j(\lambda)$ so that $q_{j+1} = \pi_j(A)q_1$ for $1 \leq j \leq i$. Set $\pi_0(\lambda) = 1$. Then,

$$\lambda[\pi_0(\lambda), \ldots, \pi_{i-1}(\lambda)] = [\pi_0(\lambda), \ldots, \pi_{i-1}(\lambda)]A_i + \beta_i \pi_i(\lambda) e_i^T,$$

or, equivalently,

$$[\pi_0(\lambda), \ldots, \pi_{i-1}(\lambda)](A_i - \lambda I) = -\beta_i \pi_i(\lambda) e_i^T. \qquad \square$$

If $r_i = \psi_i(A)r_0 \neq 0$, then the roots of the polynomial $\psi_i(\lambda)$ coincide with th eigenvalues of A_i, i.e., with the Ritz values for the projective restriction of A on $\mathcal{K}_i(r_0, A)$.

20.9   Theorem of van der Sluis and van der Vorst

Consider the ith residual's expansion

$$r_i = \sum_{j=1}^n \bar\xi_j z_j$$

in the orthonormal eigenvector basis of A. We proceed to iterate but let us launch in parallel a *comparison process* starting from the residual

$$\bar r_0 = \sum_{j=2}^n \bar\xi_j z_j.$$

Thus, there arise the vectors $\bar x_j$ generated by the comparison process together with the approximate solution vectors x_{i+j}. Set $e_{i+j} = x_{i+j} - x$ and $\bar e_j = \bar x_j - x$ (x is the exact solution of $Ax = b$).

Theorem 20.9.1 *Let $\lambda_1 \geq \ldots \geq \lambda_n$ be the eigenvalues of the matrix $A = A^* > 0$, and let θ_1 be the greatest eigenvalue of the projective restriction matrix A_i. Then*

$$\|e_{i+j}\|_A \leq c_i \|\bar e_j\|_A, \quad j = 0, 1, \ldots, \tag{20.9.1}$$

where

$$c_i = \max_{2 \leq k \leq n} \left| \frac{(\lambda_1 - \lambda_k)\theta_1}{(\theta_1 - \lambda_k)\lambda_1} \right|. \tag{20.9.2}$$

Proof. Let $r_0 = \sum_{k=1}^{n} \xi_k z_k$. Then

$$r_i = \sum_{k=1}^{n} \psi_i (\lambda_k) \xi_k z_k, \qquad r_{i+j} = \sum_{k=1}^{n} \psi_{i+j} (\lambda_k) \xi_k z_k,$$

$$\bar{r}_0 = \sum_{k=2}^{n} \psi_i (\lambda_k) \xi_k z_k, \qquad \bar{r}_j = \sum_{k=2}^{n} \psi_i (\lambda_k) \bar{\psi}_j (\lambda_k) \xi_k z_k,$$

where $\bar{\psi}_j (\lambda)$ is a polynomial of degree not higher than j.

By Lemma 20.8.1, θ_1 is the root of $\psi_i (\lambda)$, and hence, the expression

$$\Psi_i (\lambda) = \frac{1 - \frac{\lambda}{\lambda_1}}{1 - \frac{\lambda}{\theta_1}} \psi_i (\lambda)$$

is a polynomial of degree $i - 1$ satisfying the conditions $\Psi_i (0) = 1$ and $\Psi_i (\lambda_1) = 0$. Since the conjugate gradient method minimizes the A-norm of the error, we find that

$$||e_{i+j}||_A^2 = \sum_{k=1}^{n} \frac{|\psi_{i+j} (\lambda_k)|^2}{\lambda_k} |\xi_k|^2 \leq \sum_{k=2}^{n} \frac{1}{\lambda_k} |\Psi_i (\lambda_k)|^2 |\bar{\psi}_j (\lambda_k)|^2 |\xi_k|^2$$

$$\leq \max_{2 \leq k \leq n} \left| \frac{1 - \frac{\lambda_k}{\lambda_1}}{(1 - \frac{\lambda_k}{\theta_1})} \right|^2 \sum_{k=2}^{n} \frac{|\psi_i (\lambda_k)|^2}{|\lambda_k|} |\bar{\psi}_j (\lambda_k)|^2 |\xi_k|^2 = c_i ||\bar{e}_j||_A^2. \ \square$$

20.10 Preconditioning

If an iterative method applied to $Ax = b$ does not hurry to converge, then one usually attempts to apply it to some other but equivalent system $AC^{-1}y = b$. This is called *preconditioning*.

The best choice is $C = A$ (it shows, by the way, that the idea can *always* do the work). Why does no one do so?

To multiply the matrix AC^{-1} by a vector, one performs the following two actions:

- Solve a system with the coefficient matrix C;

- Multiply A by a vector.

The matrix C is said to be an *implicit preconditioner*. If one writes $AMy = b$, then M is called an *explicit preconditioner*. With explicit preconditioning, instead of solving a system with the matrix C, one carries out a matrix-vector multiplication by the matrix M. When constructing preconditioners, we keep in mind that $C \approx A$ or $M \approx A^{-1}$ (in a certain, rather broad sense).

20.11   Preconditioning for Hermitian matrices

If A is Hermitian, then AC^{-1} is generally not. However, it remains *Hermitian in a generalized sense*.

We can treat matrices as operators in a unitary space with a nonstandard scalar product, say, one of the form

$$(x, y)_D \equiv (Dx, y) \quad \text{for some} \quad D = D^* > 0$$

(check the scalar product axioms). Then, a matrix M is called *D-Hermitian* if

$$(Mx, y)_D = (x, My)_D \; \forall \, x, y \in \mathbb{C}^n,$$

and positive definite for D if $(Mx, x)_D > 0$ for all $x \neq 0$.

With such a definition, the matrix AC^{-1} is a D-Hermitian positive definite matrix with $D = C^{-1}$ or $D = A^{-1}$. Consequently, we may solve $AC^{-1}x = b$ by the standard conjugate gradient algorithm (let $D = C^{-1}$):

$$
\begin{aligned}
r_0' &= b - AC^{-1}x_0', \; p_1' = r_0'; \\
\alpha_i &= (C^{-1}r_{i-1}', r_{i-1}') \, / \, (C^{-1}AC^{-1}p_i', p_i'), \\
x_i' &= x_{i-1}' + \alpha_i p_i', \\
r_i' &= r_{i-1}' - \alpha_i AC^{-1}p_i', \\
\beta_i &= (r_i', r_i') \, / \, (r_{i-1}', r_{i-1}'), \\
p_{i+1}' &= r_i' + \beta_i p_i'.
\end{aligned}
$$

After the substitutions

$$x_i' = Cx_i, \qquad r_i' = r_i, \qquad p_i' = C^{-1}p_i,$$

the formulas turn into those of the *preconditioned conjugate gradient method*:

$$
\begin{aligned}
r_0 &= b - Ax_0, \; p_1 = C^{-1}r_0; \\
\alpha_i &= (r_{i-1}, C^{-1}r_{i-1})/(Ap_i, p_i), \\
x_i &= x_{i-1} + \alpha_i p_i, \\
r_i &= r_{i-1} - \alpha_i Ap_i, \\
\beta_i &= (r_i, C^{-1}r_i)/(r_{i-1}, C^{-1}r_{i-1}), \\
p_{i+1} &= C^{-1}r_i + \beta_i p_i.
\end{aligned}
\qquad (20.11.1)
$$

Note that x_i is an approximation to the exact solution while $r_i = b - Ax_i$ is the residual for the original system. (Derive the above formulas.)

If C and A are Hermitian positive definite matrices, then the eigenvalues of the preconditionered matrix $C^{-1}A$ are positive. (Why?) One usually tries to make them fall into a segment $[m, M]$ of length the smaller the better. The convergence theory expounded above, however, says that the convergence rate is also kept fast when most of the preconditioned eigenvalues (though not all!) are amassed near one point (this point is called a *cluster*).

Exercises

1. Let $\lambda_1 \geq \lambda_k \geq \ldots \geq \lambda_{n-k'+1} \geq \lambda_n$ be the eigenvalues of $A = A^* > 0$. Let $m' \equiv \lambda_{n-k'+1}$ and $M' \equiv \lambda_k$. Show that for the A-norm of the errors in the conjugate gradient method, the inequalities of the following form hold:

$$\|e_i\|_A \leq 2 \left(\frac{\lambda_1}{\lambda_n}\right)^{k'} \left(\frac{1 - \sqrt{\frac{m'}{M'}}}{1 + \sqrt{\frac{m'}{M'}}}\right)^{i-k-k'} \|e_0\|_A.$$

2. Let $A = A^* > 0$ be a matrix of order n, and consider the implicit preconditioner of the form $C = \text{diag}(A)$. Prove that

$$\text{cond}_2 \left(C^{-\frac{1}{2}} A C^{-\frac{1}{2}}\right) \leq n \, \text{cond}_2 \left(DAD\right)$$

for any positive definite diagonal matrix D.

Is it true that
$$\text{cond}_2 \left(AC^{-1}\right) \leq n \, \text{cond}_2 \left(AD\right)?$$

Lecture 21

21.1 Integral equations

An integral equation of the form

$$\int_\Gamma K(x,y)\, u(y)\, d\sigma(y) \;=\; f(x), \quad x \in \Gamma,$$

is said, rather formally, to be *of the first kind*, while that of the form

$$u(x) + \int_\Gamma K(x,y)\, u(y)\, d\sigma(y) \;=\; f(x), \quad x \in \Gamma,$$

is said to be *of the second kind.* Here, $d\sigma$ is the arc length element. In the operator form, we write $Ku = f$ and $(I + K)u = f$, respectively.

For definiteness, assume that $\Gamma = \{\gamma(t) : 0 \le t \le 2\pi\}$ is a sufficiently smooth closed contour on the complex plane.

The function $K(x,y)$ is termed a *kernel* of the integral equation. We assume that it is infinitely differentiable everywhere except for $x = y$, and the integrals are understood as the Riemann improper integrals.

21.2 Function spaces

It is important to consider the relevant function spaces for u and f. Since the functions u and f can be considered as 2π-periodic functions of t defined on the whole real axis, we consider the Fourier series

$$u(t) = \sum_{k=-\infty}^{\infty} u_k \, \exp(ikt),$$

and, for any real s, set

$$\|u\|_s^2 \equiv |u_0|^2 + \sum_{k \ne 0} |k|^{2s}\, |u_k|^2.$$

These quantities are defined correctly for any $u \in C^\infty$ (check this) and possess all properties of the norm.

Denote by H^s the space of which every element is identified with a set of equivalent Cauchy sequences in C^∞ with respect to the norm $\|\cdot\|_s$ (we call any two sequences equivalent if their difference converges to zero). The spaces H^s are called the *Sobolev spaces*.

Note that, for a negative s, some elements of H^s do not look like any usual function (such elements are referred to as *distributions*).

21.3   Logarithmic kernel

In the potential theory, one encounters the kernel of the form

$$K(x,y) = -\frac{1}{\pi} \ln|x-y|.$$

Let Γ be a circle of radius a. With $x = a\,e^{i\tau}$ and $y = a\,e^{it}$,

$$K(x,y) \equiv k(\tau, t) = -\frac{1}{\pi} \ln(a\,|1 - e^{i(\tau - t)}|).$$

Remember that $\ln(1-z) = -\sum\limits_{k=1}^{\infty} \frac{z^k}{k}$, for $|z| < 1$, $\Rightarrow$ With $z = e^{i\phi}$, we arrive at the expansion $\ln|1 - e^{i\phi}| = \sum\limits_{k=1}^{\infty} \frac{\cos k\phi}{k}$. Now, we obtain

$$
\begin{aligned}
[Ku](\tau) &= \frac{a}{\pi} \int\limits_0^{2\pi} \left(\ln a + \sum_{k \neq 0} \frac{e^{ik(\tau - t)}}{2\,|k|} \right) \left(\sum_m u_m\, e^{imt} \right) dt \\
&= 2a \ln a\, u_0 + a \sum_{k \neq 0} \frac{u_k}{|k|} e^{ik\tau}.
\end{aligned}
$$

This proves the following.

Proposition. *If Γ is a circle of radius $a \neq 1$, then the operator K with the logarithmic kernel provides a one-to-one continuously invertible mapping $K : H^s \to H^{s+1}$ for any real s.*

If we consider an arbitrary smooth contour, then the same holds, if not immediately, then after a small smooth perturbation of it. The proof, however, is not that straightforward.

We could regard K as an operator acting, suppose, from H^0 to H^0. Then the equation $Ku = f$ may have no solution, and, if it still has, u can change dramatically after some small perturbations of f.

21.4 Approximation, stability, convergence

For solving an operator equation $Au = f$, where $A : H \to H'$ is a continuously invertible linear operator for some Banach spaces H and H', consider a sequence of finite-dimensional projectors $P_n : H \to L_n = \operatorname{im} P_n$, $Q_n : H' \to L'_n = \operatorname{im} Q_n$, and the following *projective equation*:

$$(Q_n A P_n) u_n = Q_n f, \ u_n = P_n u_n.$$

The projective method based on P_n, Q_n is called convergent if for all sufficiently large n, the projective equations are solved uniquely and $u_n \to z$, as $n \to \infty$, for any $f \in H'$ (z is the solution to $Au = f$). For this, the following two properties are crucial:

- *Approximation property:* $P_n u \to u \ \forall u \in H, \quad Q_n f \to f \ \forall f \in H'.$

- *Stability property:* for all n sufficiently large,

$$\|(Q_n A P_n) u\|_{H'} \geq c \|P_n u\|_H \quad \forall u \in H, \quad c > 0.$$

Lemma 21.4.1 *Suppose a projective method possesses the stability property. Then the inequality*

$$\|u_n - z\|_H \leq (1 + c^{-1} \|Q_n\| \|A\|) \|z - P_n z\|_H$$

holds true for all sufficiently large n.

Proof. Subtract $Q_n A P_n z = Q_n A P_n z$ from $Q_n A P_n u_n = Q_n f$. Then $(Q_n A P_n)(u_n - P_n z) = Q_n A (z - P_n z)$. From the stability, for n large enough, $Q_n A P_n$ is an invertible mapping of $\operatorname{im} P_n$ onto $\operatorname{im} Q_n$. $\Rightarrow$

$$u_n - z = -(z - P_n z) + (Q_n A P_n)^{-1} Q_n A (z - P_n z). \ \square$$

Corollary. *With the stability property holding, if $\|Q_n\| \|z - P_n z\| \to 0$, then $u_n \to z$.*

Theorem 21.4.1 *Suppose a projective method possesses both stability and approximation properties. Then it is convergent.*

Proof. From the Banach–Steinhaus theorem, $\|Q_n\| \leq M \leq +\infty$, and it remains to refer to the above corollary. $\square$

Under the hypotheses of the above theorem, prove the following *quasi-optimality property*:

$$\|u_n - z\| \leq C \inf_{u \in L_n} \|u - z\|, \quad C > 0.$$

21.5   Galerkin method

Consider a *sesquilinear form* (v, u) defined for $v \in H'$ and $u \in H$. By definition, (v, u) is a linear continuous functional on H' for any $u \in H$ while the complex conjugate value $\overline{(v, u)}$ is a linear continuous functional on H for any $v \in H'$.

Suppose that $\phi_1, \ldots, \phi_n \in H$ and $\psi_1, \ldots, \psi_n \in H'$ are *dual* systems in the sense that $(\psi_i, \phi_j) = \delta_{ij}$ (1 for $i = j$ and 0 otherwise). In the *Galerkin method*, we set out the projectors as follows:

$$P_n u = \sum_{j=1}^{n} \phi_j \overline{(\psi_j, u)}, \quad Q_n v = \sum_{i=1}^{n} \psi_i (v, \phi_i).$$

Write $u_n = P_n u_n = \sum_{j=1}^{n} x_j \phi_j$. Then, from $Q_n A P_n u_n = \sum_{i=1}^{n} \psi_i \sum_{j=1}^{n} x_j (A\phi_j, \phi_i)$ and $Q_n f = \sum_{i=1}^{n} \psi_i (f, \psi_i)$, we derive the following Galerkin equation:

$$\sum_{j=1}^{n} (A\phi_j, \phi_i) x_j = (f, \phi_i), \quad i = 1, \ldots, n.$$

The matrix $M = [A\phi_j, \phi_i]_{n \times n}$ is sometimes referred to as the *moment matrix*.

21.6   Strong ellipticity

Given $A : H \to H'$ and a sesquilinear form on H, H', suppose that, for some $c > 0, \quad (Au, u) \geq c\|u\|_H \quad \forall u \in H$. This important property is known by different names: strong ellipticity, coerciveness, positive definiteness.

For the Galerkin method, the stability property follows immediately from the strong ellipticity. (Prove this.)

21.7   Compact perturbation

If there is a convergent projective method for $Au = f$, it can also be applied to almost any equation of the form $(A + K)u = f$, where K is a *compact* operator. Recall that a linear continuous operator $K : H \to H'$ is termed a compact operator if for any bounded sequence $u_n \in H$, a convergent subsequence can be made from the images $Ku_n \in H'$.

Theorem 21.7.1 *Suppose A and $A + K$ are continuously invertible operators from H to H', and let a projective method based on P_n, Q_n possess the approximation property and also the stability property when applied to A. Then it maintains the stability property when applied to $A + K$.*

Proof. Set $A_n = Q_n A P_n$ and $K_n = Q_n K P_n$. We take it for granted that $\|A_n u\| \geq \|P_n u\| \ \forall u \in H$. By contradiction, assume that there exists a sequence $u_n = P_n u_n$ such that

$$\|(A_n + K_n)u_n\| \to 0, \quad \|u_n\| = 1. \qquad (*)$$

Without loss of generality we may regard Ku_n as a convergent sequence. Let $Ku_n \to v \ \Rightarrow \ A^{-1}Ku_n \to A^{-1}v \ \Rightarrow$

$$A_n^{-1}K_n u_n \to A^{-1}v. \qquad (**)$$

The latter can be checked rather straightforwardly:

$$A_n^{-1}K_n u_n - A^{-1}v = (A_n^{-1}Q_n K_n u_n - A_n^{-1}Q_n v) + (A_n^{-1}Q_n v - A^{-1}v).$$

The first bracketed sequence converges to zero, because the norm $\|A_n^{-1}Q_n\|$ is bounded uniformly in n. The second converges to zero thanks to the convergence of the projective method in question, applied to $Au = v$.

From $(*)$ and due to the stability property, $(I + A_n^{-1}K_n)u_n \to 0$. With $(**)$, we infer that $u_n \to u = -A^{-1}v$. Therefore, $(A + K)u = 0 \ \Rightarrow \ u = 0$, which can not be reconciled with $\|u\| = 1$. $\quad \square$

21.8 Solution of integral equations

The stability property is guaranteed for the Galerkin method applied to the equation $Iu = f$ with the identity operator $I : H \to H' = H$. (Prove this.) Luckily enough, any second kind equation $(I + K)u = f$ with a compact operator K doe not go very far from this case. The stability property is provided by Theorem 21.7.1. Concerning the approximation property, we can take it for granted when using splines, for example.

When the solution z and the right-hand side f are sufficiently smooth, we can take care of the approximation property only on some subspaces with sufficiently smooth functions. Then the norms $\|P_n\|$ and $\|Q_n\|$ might grow, but still $u_n \to z$, whenever $\|Q_n\| \, \|P_n z - z\| \to 0$.

For the first kind equation $Au = f$, all does not seem that clear. Now, we should be more careful in the choice of relevant function spaces. Still, why not try reducing this case to the previous one?

21.9 Splitting idea

For the first kind equation $Au = f$ with a continuously invertible linear operator $A : H \to H'$, we often benefit from splitting $A = P + K$, where P is a "simple" continuously invertible operator (usually called the *principal part* of A) while K is a compact operator. All becomes simple, indeed, with P

enjoying the strong ellipticity property.

Consider, for example, the equation $Au = f$ with the logarithmic kernel (see Section 21.3). If Γ is a circle of radius $a \neq 1$, then A is a continuously invertible operator, in particular, from $H^{-\frac{1}{2}}$ to $H^{\frac{1}{2}}$. A sesquilinear form gets in naturally in the form

$$(v, u) = \int\limits_0^{2\pi} v(t)\, \overline{u(t)}\, dt.$$

The strong ellipticity property is verified straightforwardly.

In the general case, split the kernel as follows:

$$-\frac{1}{\pi} \log |x(\tau) - y(t)| = \left(-\frac{1}{\pi} \log |e^{i\tau} - e^{it}| \right) + \left(-\frac{1}{\pi} \log \left| \frac{x(\tau) - y(t)}{e^{i\tau} - e^{it}} \right| \right).$$

This induces the operator splitting $A = P + K$ with P, which we know all about, and K, which is a compact operator from $H^{-\frac{1}{2}}$ to $H^{\frac{1}{2}}$ (provided that the contour is smooth enough). Thus, we have firm grounds for applying the Galerkin method. Note that the strong ellipticity property holds not only for P, but even for A itself (after excluding some specific cases like that with a circle of radius $a = 1$). Unfortunately, I cannot suggest any short proof now.[1]

For an arbitrary continuously invertible $P : H \to H'$, we can reformulate the problem as a second kind equation $(I + P^{-1}K)u = P^{-1}f$ with the operator $I + P^{-1}K$ acting from H to H and the component $P^{-1}K$ compact.

21.10 Structured matrices

The moment matrix $M = [A\phi_j, \phi_i]_{n \times n}$ is usually *sparse* when A is a differential operator, and *dense* when A is an integral operator. By way of efficient implementation, in either case it is highly desirable to recognize a structure in it.

For example, surely, the diagonal or tridiagonal matrices are regarded as structured sparse matrices, in contrast to sparse matrices with rather chaotic zero pattern. A matrix of the form xy^T, $x, y \in \mathbb{C}^n$, is a structured dense matrices. Of course, there are many other important examples of structure.

21.11 Circulant and Toeplitz matrices

Suppose that an integral equation is given on a circle and the kernel $K(x, y)$ depends only on the distance between the points x and y (for example, the

---

[1] G. C. Hsiao and W. L. Wendland. A finite element method for some integral equations of the first kind. *J. of Math. Analysis and Appl.* 58: 449–481 (1977).

logarithmic kernel). Taking a uniform mesh $x_k = \frac{2\pi}{n}k$, $k = 1, \ldots, n$, and the piecewise constant basis functions, we discover the following structure of the moment matrix:

$$M = \begin{bmatrix} c_0 & c_{n-1} & c_{n-2} & \cdots & c_1 \\ c_1 & c_0 & c_{n-1} & \cdots & c_2 \\ c_2 & c_1 & c_0 & \cdots & c_3 \\ \cdots & \cdots & \cdots & \cdots & \cdots \\ c_{n-1} & c_{n-2} & c_{n-3} & \cdots & c_0 \end{bmatrix}.$$

Such a matrix is called a *circulant matrix*, or, in brief, a circulant.

If the contour is an arc of the circle, then, the moment matrix loses that nice structure but still captures some of its features:

$$M = \begin{bmatrix} t_0 & t_{-1} & t_{-2} & \cdots & t_{-1} \\ t_1 & t_0 & t_{-1} & \cdots & t_{-2} \\ t_2 & t_1 & t_0 & \cdots & t_{-3} \\ \cdots & \cdots & \cdots & \cdots & \cdots \\ t_{n-1} & t_{n-2} & t_{n-3} & \cdots & t_0 \end{bmatrix}.$$

The entries of M along the line $i - j = k$ are the same. The matrix is totally determined by the entries of its first column and row. Such a matrix is termed a *Toeplitz matrix*.

21.12 Circulants and Fourier matrices

Structures can provide tremendous savings in memory and arithmetic costs. Instead of n^2, circulants and Toeplitz matrices can be stored in n and $2n - 1$ memory cells, respectively. Low arithmetic costs for them are accounted for by a profound relation between circulants and *Fourier matrices* of the form

$$F_n = [w^{kl}], \quad 0 \le k, l \le n - 1, \quad w = e^{-i\frac{2\pi}{n}}.$$

Theorem 21.12.1 *Suppose that $C \in \mathbb{C}^{n \times n}$ is a circulant with the first column $c \in \mathbb{C}^n$. Then*

$$C = \frac{1}{n} F_n^* \operatorname{diag}(F_n c) F_n.$$

Proof. Let $\zeta = w^k$, and set

$$\lambda_k = \sum_{j=0}^{n-1} \zeta^j c_j \quad \Rightarrow \quad \zeta^l \lambda_k = \sum_{j=0}^{n-1} \zeta^j c_{j-l \,(\text{mod } n)}.$$

Gathering these equations for all l and k, we arrive at $F_n C = \operatorname{diag}(F_n c) F_n$. It remains to verify that $F_n^{-1} = \frac{1}{n} F_n^*$. $\square$

One multiplies a circulant by a vector through multiplying the Fourier matrix three times by a vector. The same applies to solving a linear system with a circulant coefficient matrix.

To multiply a Toeplitz matrix T_n by a vector, we reduce the problem to that with a circulant matrix C by imbedding T_n in C:

$$C = \begin{bmatrix} T_n & \cdots \\ \cdots & \cdots \end{bmatrix} \in \mathbb{C}^{N \times N}.$$

This can be done for any $N \geq 2n - 1$. (Prove this!)

21.13 Fast Fourier transform

It is possible to compute $y = F_n x$ in $O(n \log n)$ operations. Some people say that this goes back to Gauss. We are aware now that Runge and Lanczos knew this. At any rate, a boom in activities around the fact was triggered off in 1965 by J. W. Cooley and J. W. Tukey.[2] We outline here the idea for doing this.

Let $n = 2m$, and denote by P_n the permutational matrix obtained from I by ordering the rows as follows: $1, 3, \ldots, m, 2, 4, \ldots, 2m$. Then it is not difficult to verify that

$$P_n F_n = \begin{bmatrix} [w^{2kl}] & [w^{2k(m+l)}] \\ [w^{(2k+1)l}] & [w^{(2k+1)(m+l)}] \end{bmatrix}, \qquad 0 \leq k, l \leq m - 1. \qquad \Rightarrow$$

$$P_n F_n = \begin{bmatrix} F_m & 0 \\ 0 & F_m \end{bmatrix} \begin{bmatrix} I_m & 0 \\ 0 & W_m \end{bmatrix} \begin{bmatrix} I_m & I_m \\ I_m & -I_m \end{bmatrix}, \qquad (21.13.1)$$

where $W_m = \operatorname{diag}\{w^0, w^1, \ldots, w^{m-1}\}$.

Thus, we sort out the problem with F_{2m} by reducing it to the two problems with F_m. If n is a power of 2, then we need $\frac{1}{2} n \log_2 n$ complex multiplications and $n \log_2 n$ complex addition-subtraction operations. (Check this!)

21.14 Circulant preconditioners

Consider an integral equation with the logarithmic kernel on a smooth closed contour $\Gamma = \{\gamma(t) : 0 \leq t \leq 2\pi\}$. Using a uniform mesh on $[0, 2\pi]$ and bringing in a new unknown function in the form $U(t) \equiv u(t) |\gamma'(t)|$, we split the moment matrix $M = C + R$ so that C, corresponding to the principal part of the operator, is a circulant matrix. Since $C^{-1}R$ can be thought about as a discrete analog of a compact operator, C is anticipated to be a good

---

[2] J. W. Cooley and J. W. Tukey. An algorithm for the machine calculation of complex Fourier series. *Math. Comput.* 19 (90): 297–301 (1965).

preconditioner for M, because the eigenvalues of $C^{-1}M$ are clustered at 1.

We may look for a circulant preconditioner for M without reference to the operator splitting. For example, we can follow the recent advice of T. Chan[3] and take an *optimal circulant* C, one that minimizes $||M - C||_F$ over all circulants C.[4]

Circulants prove to be especially efficient for preconditioning Toeplitz matrices. Let $A_n = a_{i-j} \in \mathbb{C}^{n \times n}$ be a sequence of Toeplitz matrices, and let C_n be the corresponding optimal circulants.

Proposition. Suppose that $\sum_{k=-\infty}^{\infty} |a_k|^2 < +\infty$. Then $||A_n - C_n||_F^2 = o(n)$.

This can be verified by a direct calculation.

Assume that C_n are invertible. If the norms $||C_n^{-1}||_2$ are uniformly bounded in n or grow so that $||C_n^{-1}||_2 ||A_n - C_n||_F = o(\sqrt{n})$, then

$$||I_n - C_n^{-1} A_n||_F = o(\sqrt{n}).$$

Now it follows from the results presented in Lecture 5 that the singular values of $I_n - C_n^{-1} A_n$ are clustered at 0. From the same lecture, we know that, in general, this causes the eigenvalues of $I_n - C_n^{-1} A_n$ to be clustered at 0. If so, then the eigen and singular values of $C_n^{-1} A_n$ are clustered at 1 (prove this).

Exercises

1. Suppose that Γ is a circle of radius $a \neq 1$. Prove that the operator K with the logarithmic kernel provides a one-to-one continuously invertible mapping $K : H^s \to H^{s+1}$ for any real s.

2. Suppose that a projective method with projectors P_n, Q_n possesses both approximation and stability properties, and $A : H \to H'$ is a continuously invertible linear operator. Prove that the sequence of approximate solutions u_n converges to an exact solution z quasi-optimally, that is,

$$||u_n - z|| \leq C \inf_{u \in L_n} ||u - z||, \quad C > 0, \quad L_n = \text{im } P_n.$$

3. For the Galerkin method, prove that the strong ellipticity property implies the stability property.

---

[3]T. Chan. An optimal circulant preconditioner for Toeplitz systems. *SIAM J. Sci. Statist. Comput.* 9: 766–771 (1988).

[4]For other approaches, see, for example, R. Chan and G. Strang. Toeplitz equations constructed by conjugate gradients with circulant preconditioners. *SIAM J. Stat. Comput.* 10: 104–119 (1989). Note that there are many other (quite recent) works.

4. Let $k(\tau, t)$ be a continuous function of τ and t, and consider the following integral operator:

$$[Ku](\tau) \;=\; \int_0^{2\pi} k(\tau, t)\, u(t)\, dt.$$

Prove that $K : C[0, 2\pi] \to C[0, 2\pi]$ is a compact operator.

5. Under the notation of the previous problem, consider the projective method for the equation $(I + K)u = f$, $u, f \in C[0, 2\pi]$, with $P_n = Q_n$ being the interpolative projector on the Chebyshev nodes. Prove that $\|u_n - z\|_{C[0,2\pi]} \to 0$, provided that z is a continuously differentiable function.

6. Prove that if the eigenvalues of $A_n \in \mathbb{C}^{n \times n}$ are clustered at 0, then those of $I_n + A_n$ are clustered at 1.

7. Consider an integral operator

$$[Ku](x) \;=\; \int_0^1 K(x, y)\, u(y)\, dy$$

with the kernel $K(x, y)$ a continuous function of x and y. Let M_n be the moment matrices corresponding to the piecewise constant basis functions on uniform meshes (with n nodes). Prove that the singular and eigen values of M_n are clustered at zero. Is the cluster proper?

8. In the previous problem, do the eigenvalues of M_n approximate the eigenvalues of the operator K?

9. Let F_n be the Fourier matrix of order n. Prove that $n^{-1/2}\, F_n$ is a unitary matrix.

10. Prove the equality (21.13.1).

11. Let $A_n = [a_{i-j}]_{n \times n}$ be a Toeplitz matrix and C_n be the corresponding optimal circulant. Denote by c_k^n, $k = 0, 1, \ldots, n-1$, the entries of the first column of C_n. Prove the following formulas:

$$c_k^n \;=\; \sum_{i,j:\; i-j=k \;(mod\,n)} a_{i-j} \;=\; \frac{(n-k)\,a_k + k\,a_{n-k}}{n}.$$

12. Prove the proposition from Section 21.14.

13. Assume that $A_n = A_n^* > 0$. Prove that if C_n is the optimal circulant for A_n, then $C_n = C_n^* > 0$.

Bibliography

K. I. Babenko. *Fundamentals of Numerical Analysis*. Nauka, Moscow, 1986 (in Russian).

N. S. Bakhvalov, N. P. Zidkov, and G. M. Kobelkov. *Numerical Methods*. Nauka, Moscow, 1987 (in Russian).

I. S. Berezin and N. P. Zidkov. *Methods of Computations*. Moscow, 1962 (in Russian).

E. G. D'yakonov. *Minimization of Computational Work*. Nauka, Moscow, 1989 (in Russian).

D. K. Faddeev and V. N. Faddeeva. *Computational Methods of Linear Algebra*. San Francisco–London, 1963.

G. E. Forsythe, M. A. Malcolm, and C. B. Moler. *Computer Methods for Mathematical Computations*. Prentice-Hall, Inc., Englewood Cliffs, N. J., 1977.

A. O. Gelfand. *Calculation of Finite Differences*. Moscow, 1952 (in Russian).

S. K. Godunov. *Solution of Systems of Linear Equations*. Nauka, Novosibirsk, 1980 (in Russian).

G. H. Golub and C. F. Van Loan. *Matrix Computations*. The Johns Hopkins University Press, Baltimore, 1989.

R. A. Horn and C. R. Johnson. *Matrix Analysis*. Cambridge University Press, Cambridge, England, 1986.

Kh. D. Ikramov. *Nonsymmetric Eigenvalue Problem*. Nauka, Moscow, 1991 (in Russian).

198

Kh. D. Ikramov. *Numerical Methods for Symmetric Linear Systems.* Nauka, Moscow, 1988 (in Russian).

V. P. Il'in and Yu. I. Kuznetsov. *Algebraic Bases of Numerical Analysis.* Nauka, Novosibirsk, 1986 (in Russian).

V. G. Karmanov. *Mathematical programming.* Nauka, Moscow, 1975 (in Russian).

G. I. Marchuk. *Methods of Numerical Mathematics.* Nauka, Moscow, 1989 (in Russian).

J. M. Ortega and W. C. Rheinboldt. *Iterative Solution of Nonlinear Equations in Several Variables.* University of Maryland, College Park, Maryland, 1970.

A. M. Ostrowski. *Solution of Equations and Systems of Equations.* Academic Press, New York, London, 1960.

B. N. Parlett. *The Symmetric Eigenvalue Problem.* Prentice-Hall, Inc., Englewood Cliffs, N. J., 1980.

S. Prössdorf and B. Silbermann. *Numerical Analysis for Integral and Related Operator Equations.* Birkhauser, Boston, 1991.

A. A. Samarski, A. V. Gulin. *Numerical Methods.* Nauka, Moscow, 1989 (in Russian).

G. W. Stewart and J. Sun. *Matrix Perturbation Thery.* Academic Press, San Diego, 1990.

A. G. Sukharev, A. V. Timokhov, and V. V. Fedorov. *A Course of Minimization Methods.* Nauka, Moscow, 1986 (in Russian).

J. F. Traub, G. W. Wasilkowski, and H. Wozniakowski. *Information-Based Complexity.* Academic Press, San Diego, 1988.

V. V. Voevodin. *Computational Bases of Linear Algebra.* Nauka, Moscow, 1977 (in Russian).

J. H. Wilkinson. *The Algebraic Eigenvalue Problem.* Clarendon Press, Oxford, 1965.

Yu. S. Zav'yalov, B. I. Kvasov, and V. L. Miroshnichenko. *Methods of Spline-Functions.* Nauka, Moscow, 1980 (in Russian).

Index